Do it yourself GUIDE TO GLASS, TIMBER AND TILES

Do it yourself GUIDE TO GLASS, TIMBER AND TILES

GLASS AND GLAZING · WOOD AND WOOD FINISHING
TILES AND TILING

BLANDFORD PRESS
POOLE · DORSET

First published in the UK 1985 by
Blandford Press, Link House, West
Street, Poole, Dorset, BH15 1LL.

Distributed in the United States by
Sterling Publishing Co., Inc., 2 Park
Avenue, New York, N.Y. 10016.

British Library Cataloguing in Publication Data

The do it yourself guide to glass, timber and tiles
1. Dwellings—Remodeling—Amateurs' manuals
I. Do it yourself
643'.7 TH4816

ISBN 0 7137 1512 X

Printed in Hong Kong by South
China Printing Co.

Acknowledgements

The publishers gratefully acknowledge the following for use of illustrations and products.

AG Tiles p.85 (6), Contiboard p.43, Copydex p.56 (4), Roy Day p.21 (7), Dunlop p.86, Everest p.13 (photo 11), p.27 (11), Garfield Glass p.75 (5&9), James Halsted p.85 (2&4), Heuga Tiles p.74 (1), p.84, 85 (5), H & R Johnson p.72 (2), p.94, KEF p.58 (1), Magnet Southern p.50 (2&5), p.51 (8), p.52 (11&12), p.53, Marley p.73, p91-93, Plannja International p.72, Polycell p.26, p.27 (13), Redland p.69, Siesta Tiles p.74 (3), p.83, Velux p.18 (photo), Vencel p.74 (4), p.75, Vigers p.40 (9-11), p.51 (6&7), Wicanders p.85 (1).

GLASS AND GLAZING

Contents

Looking glass

Glass has a history going back some 4,000 years. And we've come a long way with this versatile material since Phoenician sailors accidentally made glass while cooking over a wood fire on a sandy beach.

The Egyptians made jewellery of it. Solomon used it as floor tiles in his palace. The medieval cathedrals recounted stories to folk who couldn't read by their wonderful stained glass windows, while the Palace of Versailles would be a duller place without its hall of mirrors.

Today, unfortunately, we very much take glass for granted. All our homes are glazed with it, and we've learned to look through it rather than at it as a decorative material with vast possibilities. Yet glass comes in a bewildering array of patterns, colours and textures which can transform the appearance of a home or a room.

Apart from glazing, it can be used for screens and partitions, doors, mirrors, table tops and shelves. And it has a role to play in double glazing and in home protection, while glass blocks can be used to let light into halls and passageways.

THINK SAFETY!

If used incorrectly glass can become a hazard. But unlike most home accidents, those with glass can be eliminated if the correct materials are used. Glazed front and back doors, french windows – especially if panels are child-height – should be glazed with safety glass. And the same applies to doors that could slam shut in high winds. And all low silled windows should be fitted with safety glass – and the same applies to infill panels, such as staircases and landings or balconies.

So, before you buy ... think!

Patterns and textures

Glass doesn't have to be plain. A very good range of patterns and textures is available, some of which are illustrated here, taken from the Pilkington patterned glass range.

1 Hammered
2 Mayflower
3 Patchwork
4 Pimpernel
5 Reeded
6 Sycamore

1

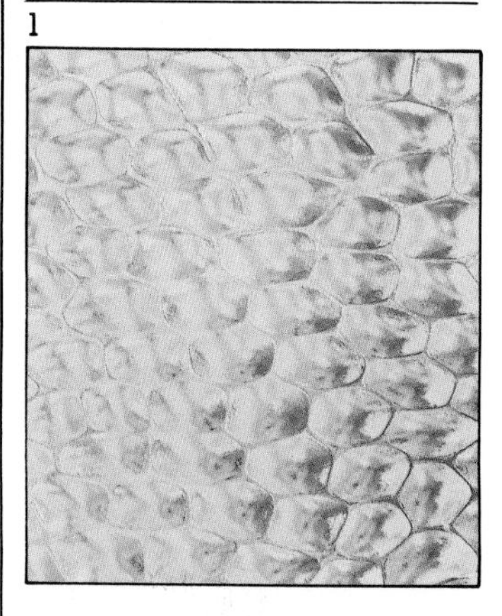

3

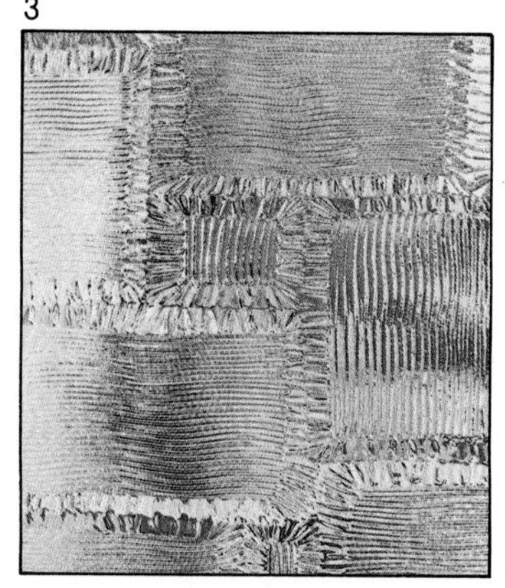

5

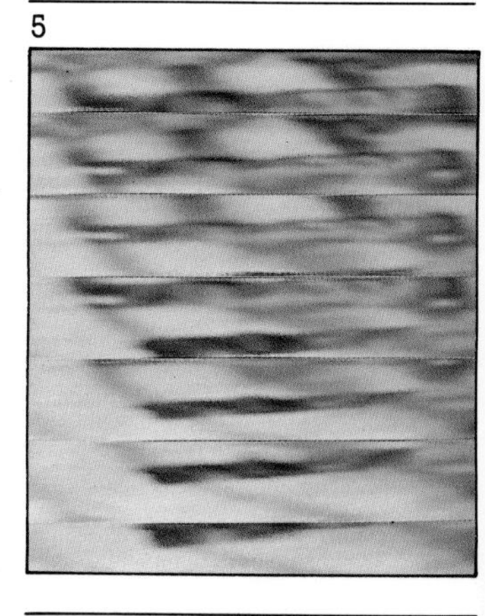

2

4

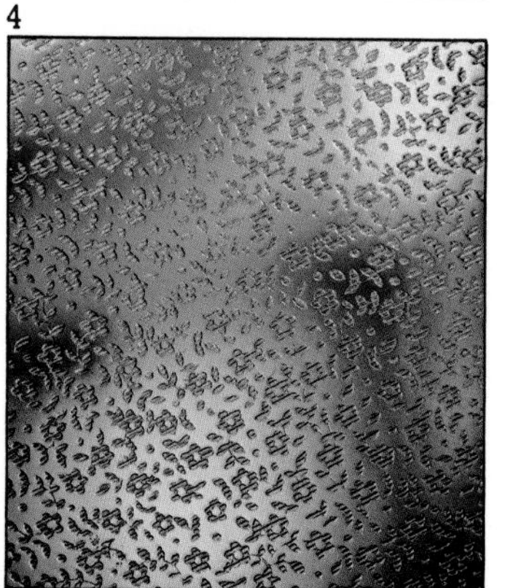

6

7

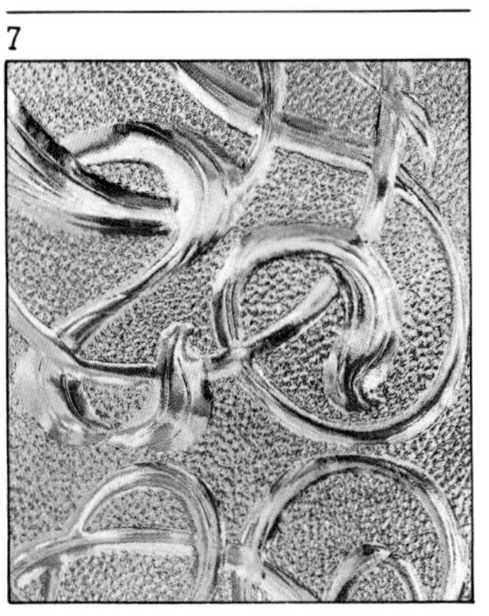

9

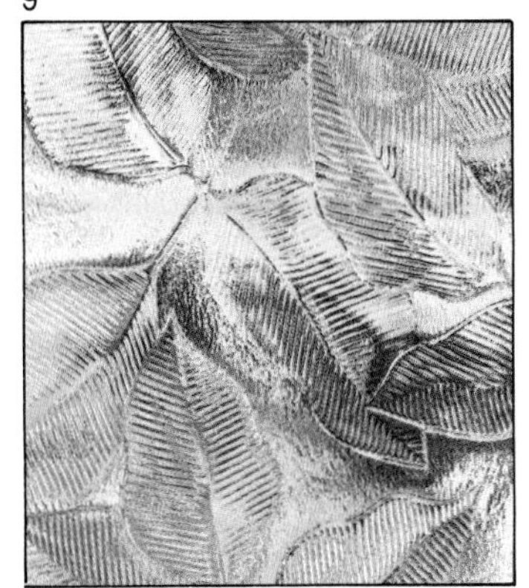

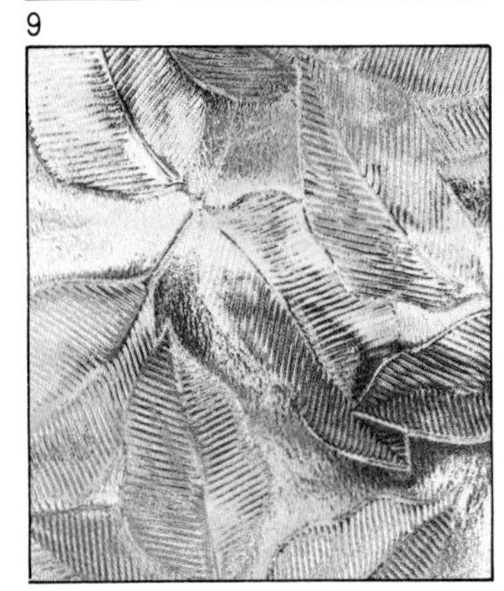

11

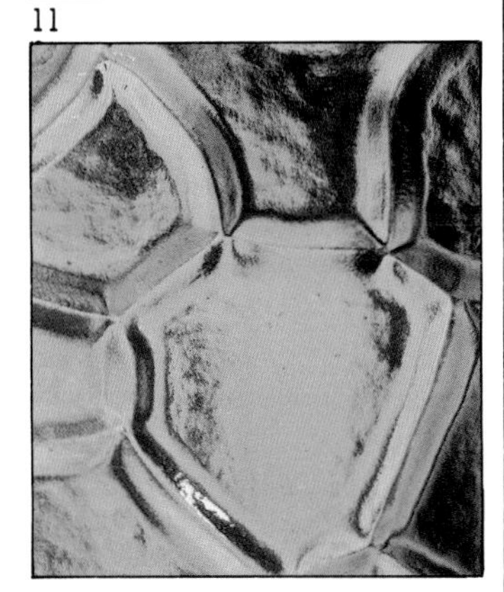

8

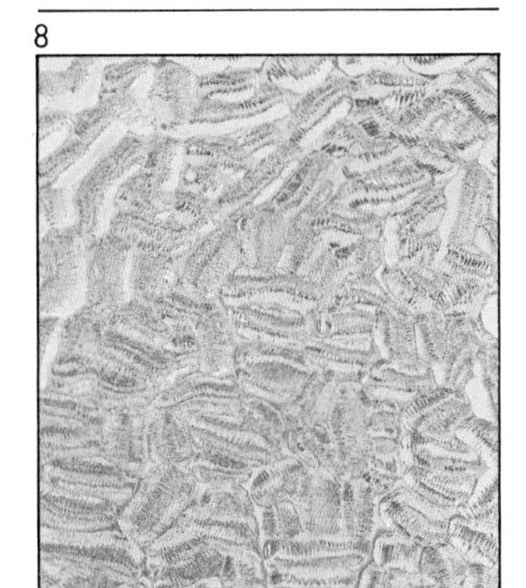

10

12

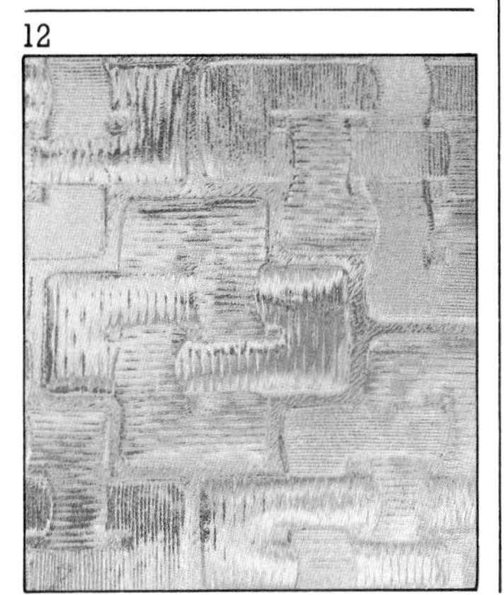

7 Everglade
8 Arctic
9 Autumn
10 Driftwood
11 Deep Flemish
12 Linkon

Patterns and textures

1 Ordinary glass breaks into dangerous pieces
2 Glass protected by a film of plastic holds together, and considerable force is needed even to make a hole

ORDERING GLASS

It is essential that measurements are accurate when ordering glass, for trimming away thin slivers is not easy for the amateur. For any window glass, allow about ⅛in. (3mm) less on each dimension so your bedding putty will cushion the glass in the frame. And it is wise to check diagonals on older frames to make sure the actual openings are true.

Where you encounter odd shapes, make a pattern and take this to the glazier. Further details of cutting your own glass, and of patterns can be found on pages 4 and 24.

SPECIAL GLASS

Apart from the variety of patterns illustrated on the previous pages, a number of special glasses are available for domestic use. For example:

Wired glass
This may be clear or obscured glass with a wire mesh sandwiched between. Used for surfaces like the roofs of sun lounges, it affords greater strength and, should the glass be shattered, the pieces are held together by the wire. Useful as a fire barrier, but not much protetion against break-ins.

Toughened glass
Set sizes of glass are specially treated to greatly increase the impact resistance of the glass. Ideal for glass doors. And if the material should be broken, it shatters into hundreds of harmless pieces, with no jagged edges to cause injury. This glass cannot be cut to size once it has been toughened.

Laminated glass
Clear glass is used either side of a thin film of very tough plastic which becomes invisible once the sheets are welded together by heat. The material is very tough, and while impact will crack the glass, the film holds it all very firmly together. Ideal as safety glass, and provides excellent security against forced entry. May be cut to size any time.

Your glazier will be pleased to advise you on other specials, such as non-reflective glass for picture frames; prismatic glass for deflecting light into a dark area and bullion glass for enhancing the character of period windows and cottages.

He can also quote for special jobs, like smoothing the edges of cut mirror, table tops and glass shelves.

Which glass where?

Decisions, decisions! As you've probably gathered, there are many and varied uses of glass around the home, and in the following pages you can see just a few attractive uses of glass. Here are a few guidelines:

Windows

Check on page 34 for recommended thickness for given size windows. Remember that windows which might be ideal for a thief to break to reach window catches can be fitted with laminated glass. For windows like bathrooms and loos, remember there is a large range of patterns in obscured glass. Always fix with pattern side inwards so you get the smoothest putty seal on the outside.

Where new windows are to be fitted, frames with space for sealed double glazed units can be used. This is obviously the

1

2

3

4

1 Toughened glass used as a pool windbreak
2 The glazed door blends well with the attractive bow window
3 A neat lean-to greenhouse for where space is limited
4 Large areas of glass help in home extensions where light must pass to the adjoining lounge

Which glass where?

5 A vital area for toughened or laminated glass (covered by BS 6713, balconies)
6 Lots of light from this combined window and doorway in patterned glass. Doors come toughened.
7 An example of how colour and pattern in glass can become part of the décor. The doors are a feature in their own right

cheapest time to buy.

Doors.
In any risk areas, especially where children are about, use safety glass in glazed areas. Where you want extra security, such as with a front door with glazed panel, fit laminated glass held in place by a stout beading.

To improve internal daylight, replace existing solid doors with toughened glass ones. They look attractive, and tend to reflect the decor of a room. Good for letting light into dingy passages.

Greenhouses and workshops
Remember that horticultural glass is cheaper for areas where perfect clarity is not required. Also useful for cloches and frames.

Shower doors and screens.
Remember to use a toughened glass in such locations. A firm frame is also required to hold the glass.

5

6

7

8

9

10

11

12

13

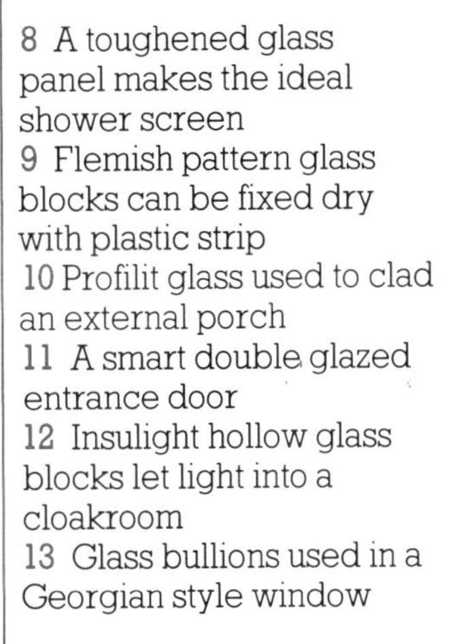

8 A toughened glass panel makes the ideal shower screen
9 Flemish pattern glass blocks can be fixed dry with plastic strip
10 Profilit glass used to clad an external porch
11 A smart double glazed entrance door
12 Insulight hollow glass blocks let light into a cloakroom
13 Glass bullions used in a Georgian style window

Which glass where?

Where glazing is concerned, areas of risk are covered by British Standard Code of Practice BS 6262. It recommends that the glazing in a fully glazed door or patio door should be of a safety glazing material to BS 6206. Doors with more than one pane where a single pane does not take up most of the door, it can be glazed in 6mm or thicker normal glass. Where a door side panel could be mistaken for a door, the same recommendations apply.

With regard to low level glazing, the code recommends that in places where many people, especially children, are moving about or where glazing comes lower than 31in (800mm), safety glazing should be used. Where glazing is protected by a barrier rail, annealed glass can be used.

The illustration below shows glazed doors and side panels which could be mistaken for doors

Not less than 6mm annealed glass.

Safety glazing material

See also the table of glass thicknesses on page 34

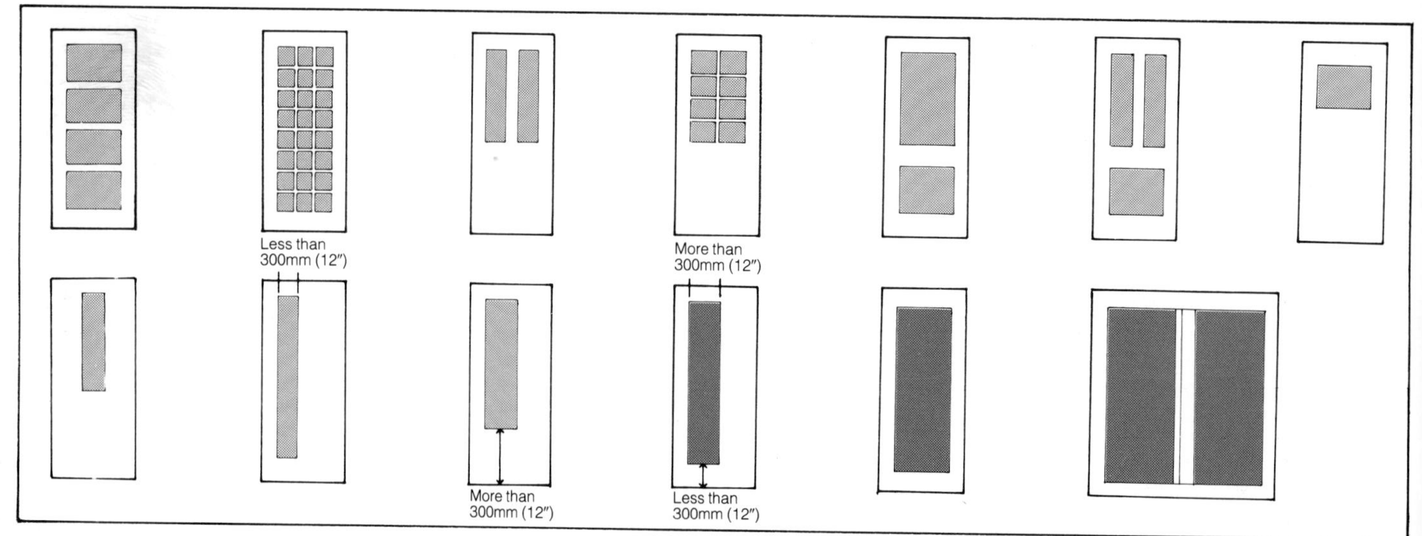

The problem of noise

Type of Glazing	Mean noise reduction in decibels
Single Glazing	
.4mm annealed glass	25
.6mm " "	27
.8mm " "	29
.10mm " "	30
.12mm " "	31
.6mm laminated glass	29
.12/14mm " "	34
Double Glazing	
.6/12/6* unit	29
.10/12/6 unit	31
.10/100/6 unit	39
.6/200/6 unit	41
.6/200/6 with sound absorbency in reveals	44

Decibel Change	Effect
3	Just noticeable
5	Quite noticeable
10	Twice as loud (or soft)
20	Much louder (or softer)

*Thickness of Glass / Airspace / Glass in millimetres

Decibel rating	Typical Sounds
120	Jet aircraft at 500ft. or inside boiler factory.
110	Motor horn at 20 ft. Pop group at 4ft. Power mower at 4ft.
100	Food blender at 2ft. Train door slamming.
90	Heavy truck. Underground train.
80	Inside small car. Noisy office.
70	Rush hour traffic. Building noise.
60	Loud conversation at 3ft.
50	Quiet street. Inside average home.
40	Quiet office. Quiet conversation. Residential area at night.
30	Tick of watch. Rustle of paper. Whisper.
20	Quiet country lane.
10	Leaves rustling in the wind.
0	Threshold of hearing.

While noise from neighbours is often a problem, many noise problems come from outside the house, varying from passing traffic to aircraft passing overhead. And ordinary single glazed windows can often be a weak link in your home's defences against external noise.

First, deal with gaps and cracks around windows, for it is estimated that if the area of gaps is just 1% of the total window area, you can lose 10dB of noise insulation. To get an idea of the different levels of noise, see the chart of typical sounds. Scientifically, noise is measured in decibels, with 0dB being the threshold of hearing where we can just discern a sound. When you get up to 120dB it can actually cause a pain.

Thicker glass will improve sound insulation, as is illustrated in the table. And laminated glass further adds to the insulative effect because of its flexible plastic interlayer which gives a dampening effect to the noise waves.

When glass gets broken

1. Check diagonals to see frame is true
2. **Wiggle out all broken glass**
3. Chisel out any remaining putty
4. Dust out all remaining debris
5. Prime all bare wood to stop suction
6. Apply bedding putty to the rebate
7. Always press round the edges of the glass
8. Slide the hammer across the glass when pinning
9. Apply the surface putty,

1

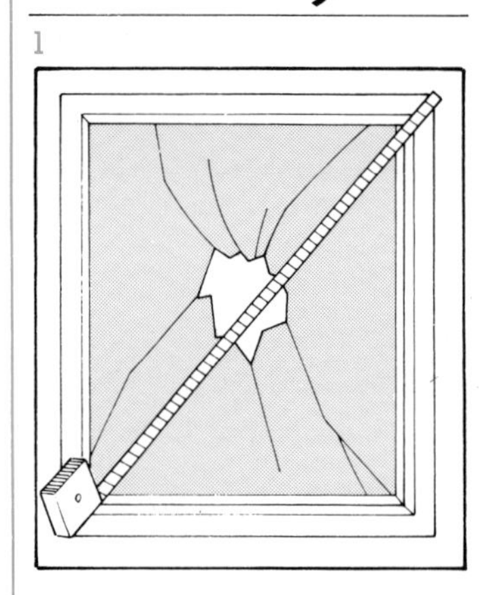

2

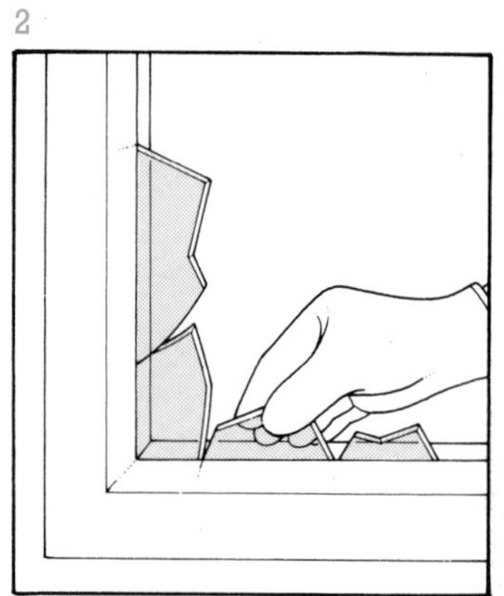

3

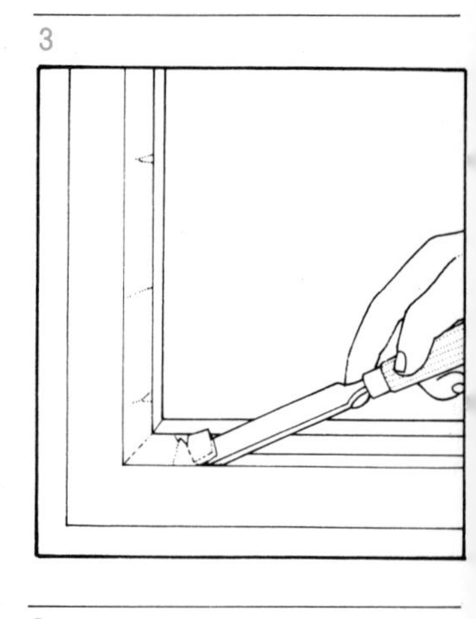

6

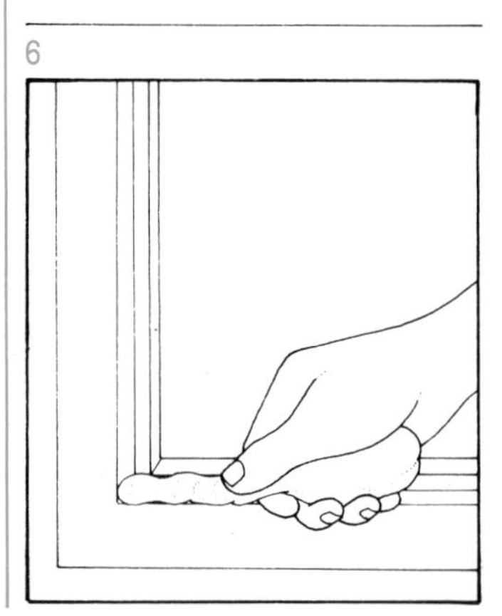

7

8

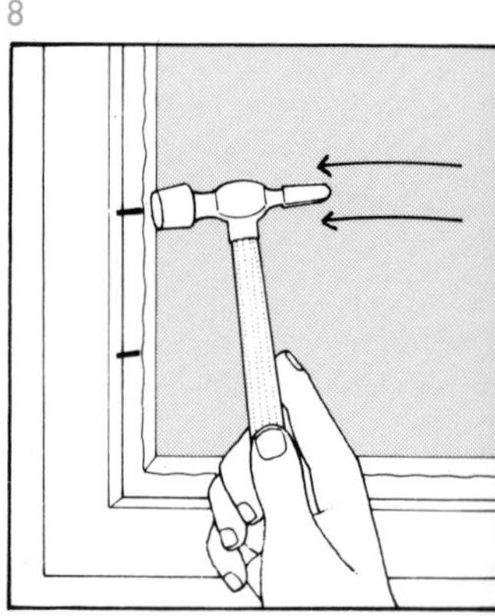

EMERGENCY

Where a pane of glass is cracked but no glass has fallen out, as a temporary measure seal the cracks with waterproof glazing tape. This is a transparent self-adhesive tape which will keep the weather out until the glass can be replaced. Don't use standard cellophane tape. It will break down if it gets wet.

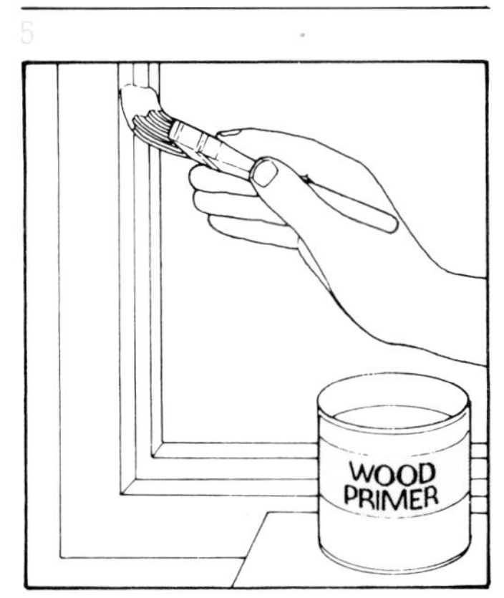

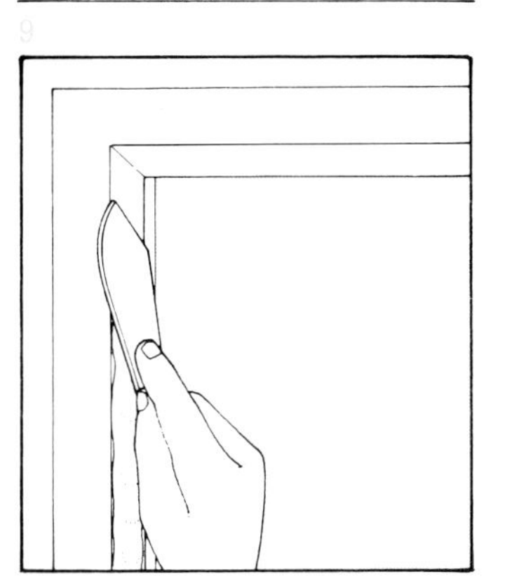

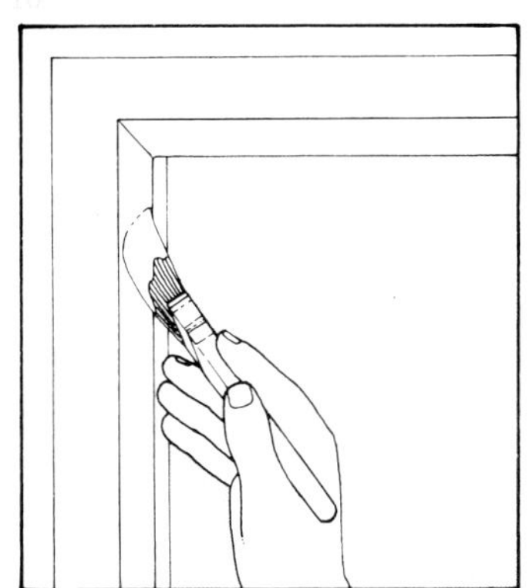

Before ordering a replacement piece of glass for a broken window, check whether the thickness is adequate for the location. See ordering glass, page 10, and risk areas page 12. If the glass was broken to force entry, consider using a laminated glass for replacement. This would resist all but the most determined villain.

Remove all broken glass from the frame, and if possible wear a tough leather gardening glove to protect your hand. Chip out all remaining putty with an old chisel kept for this kind of job and a mallet. It is wise to wear safety specs.

Brush the frame clean, then treat the rebate with wood primer if a timber frame to prevent the wood drawing the oil from your new putty. Metal frames need no treatment unless there are signs of rust. In which case use a rust inhibitor.

You need linseed oil putty for timber frames, or metal casement putty for metal frames. Enough to form a bed for the new glass plus the new bevelled finish.

Lay a bed of putty in the rebate, and press the glass into it so the putty is squeezed out around the glass. This will cushion it against severe vibration.

The final bevel of putty is the trickest part for the inexperienced, and there are a couple of simple devices on the market which help get the bevel more neatly than a putty knife. It is important that the putty be pliable, and in cold weather it pays to work the putty in your hands to soften it. When putty contains too much oil, roll the putty in newspaper before use.

Patterns for awkward shapes

1 Remember to mark which side the glass texture is on your pattern
2 Also make it clear which way the pattern is to run

1

2

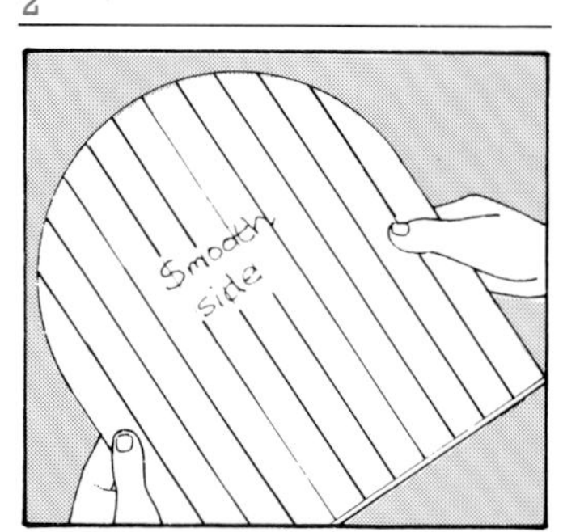

Where you have to replace a piece of glass which has a shape impossible to order by measure, make a simple paper pattern using stiffish brown paper or thin card. Remember you want this to be ⅛in. (3mm) smaller all round so there is room for the bedding putty to squeeze around the glass.

Solar control

This attractive window would be ideal for solar control glazing

Obviously, we want our windows to let in the daylight. But during summer months, South-facing windows can act as an excellent sun trap making rooms far too hot for comfort.

Solar Control glasses help solve this problem by reflecting and absorbing the sun's heat to reduce the amount coming through.

This type of glass can be very effective, cutting down the sun's heat entering a room by as much as 90%. Another benefit, of particular interest to the housewife, is their anti-fade properties. By reducing strong sunlight they cut down the amount of fading in curtain fabrics and furnishings.

The newer reflecting glass offers privacy without the drawing of heavy curtains, while from the outside, the mirrored surface tends to reflect the garden and surrounds. And there is now a laminated solar control glass which has a specially tinted inter-layer. This also acts as a security barrier where required.

If you want to treat existing glass, thin plastic films are now available which can be stuck to any clean smooth surface. Once dry, the film is crystal clear, while adding the properties of a solar control coating. As with laminated glass, the film also has a toughening effect.

Mirrors for all occasions

While d-i-y home-improvers constantly amaze us with their seemingly unlimited imagination, resourcefulness and supply of new ideas, it is surprising to learn that recent research on the use of mirrors in the home has unearthed something of a blind spot. It appears that up until quite recently, little thought has been given to the wide range of practical advantages which can be derived from the clever and subtle use of mirrors around the house. In this respect we are well behind current thinking and practices on the Continent and in the United States.

MAKING SPACE

The day has long gone when mirrors functioned simply as looking glasses. Nowadays, they should be regarded as much a part of home-improvement planning as the selection of wallpaper or carpeting.

We often hear householders complain that their homes lack a sense of space; that some of the rooms have a claustrophobic effect; or that the lighting is inadequate. It is in these areas of space and lighting that practical and artistic d-i-y skills involving the use of mirrors can produce really outstanding results. For example, simply covering one wall of a room with mirrors immediately creates the illusion of a doubling its size.

By the same token, mirrors can be particularly effective when used to compensate for disproportionate room sizes – especially with oblong-shaped living rooms, through lounges, small bedrooms and narrow hallways. By merely mounting mirrors, floor-to-ceiling on the two long walls, extending back 600mm to 1200mm from the angle formed with the short wall, a feeling of much greater width can be achieved.

GARDEN VIEW

This can appear quite striking if the short wall carries floor-to-ceiling curtains over its full width. Again, in rooms with patio doors and Venetian blinds, mirrors can give additional breadth to the view of the garden.

If floor-to-ceiling mirrors seem extravagant, an alternative choice is to use two rows of 600mm high mirrors mounted along the wall – the first about 600mm above the floor, and the second about 150mm above the first. Much the same effect will be achieved.

Mirrors can be used to great advantage in some of the smaller rooms of the house. For example, a rather cramped bathroom can

seemingly be transformed overnight through the contrived effects of mirrors to produce a feeling of more space. This can be achieved either by covering a whole wall with mirrors, or just the top half of the wall if bathroom tiling already reached half way up.

Another creative ploy is to place mirrors behind statues, possibly in an alcove, and behind other decorative objects, such as trophies,

Mirrors for all occasions

1 A kitchen 'enlarged' by the use of five mirror panels
2 Mirror neatly inset into a shelving unit
3 Glass-fronted wardrobes add an air of spaciousness to a modern bedroom

1

2

3

4

5

6

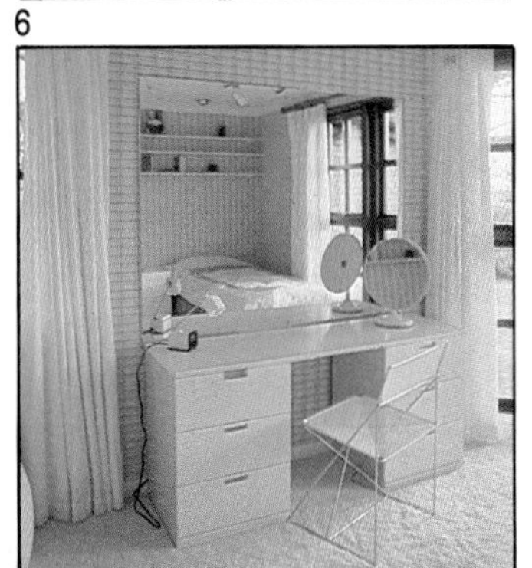

7

8

4 Large mirrors behind basins give depth to a bathroom

5 Mirror panels used as an alcove feature

6 Dressing table enhanced by a full width mirror

7 TV personality Roy Day uses mirror tiles to good effect

8 Mirror panels on both walls create an 'endless' bathroom

Mirrors for all occasions

9 Mirrors fixed to hinged timber doors, and running floor to ceiling, apparently double the size of the room
10 An old hallway badly needing a facelift
11 The same hall after decorating and the addition of a wall of mirror

family heirlooms, ornamental lamps and shelf-units. Particularly effective is a mirror sited behind a floral display, for this can produce an enchanting splash of reflected colour.

SIMPLE FIXING

The reasonably competent do-it-yourselfer will find little difficulty in the business of fixing mirrors. One simple method for mirrors up to about 1220mm by 915mm is to use special mirror screws which have a chrome-plated dome cover to hide the slotted screw heads after fitting. The screws go through holes drilled in the mirror at least 25mm from the edge, and should be driven into suitable plugs for solid or cavity walls. Behind each mirror screw should be a nylon washer to provide air and circulation space, and to cushion the mirror itself.

9

10

11

Problems may be encountered when fixing mirrors to very uneven walls. However the difficulty can be overcome by fixing battens to the wall surface and screwing the mirror to the battens after they have been carefully levelled. It is important to provide adequate cross-battening to support the mirror between fixing points, as pressure on unsupported areas could cause it to break.

Very large mirrors can be supported by special corner plates, though the plates and the visible screws holding them may be considered unsuitable for certain styles of décor. An alternative method is to use mirror clips, which are simply screwed to the wall, and hook or slide into position around the edge of the mirror.

The fixing or mirror tiles presents no special problem other than to ensure that only clean, dry and reasonably flat surfaces are used as a base. Porous surfaces, such as plaster, emulsion paint, wood, plywood, chipboard and hardboard should first be well sealed with gloss paint only (not vinyl). Wall-

continued on page 31

1

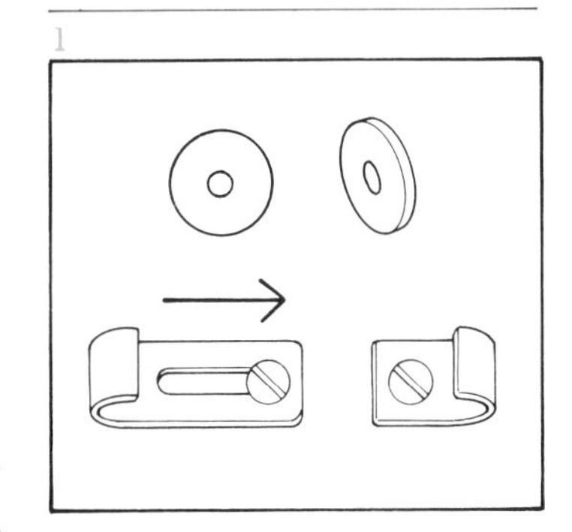

2

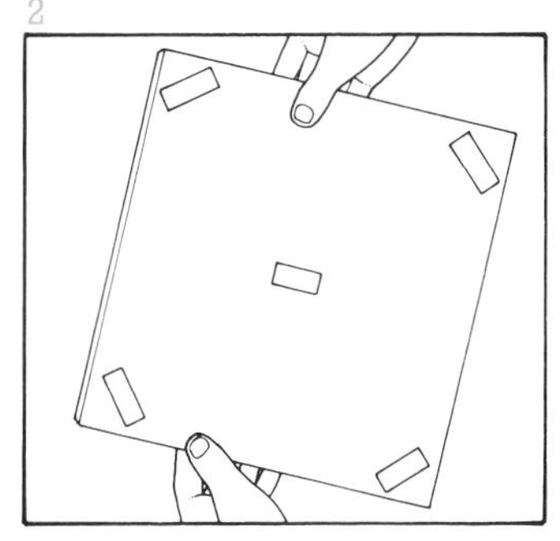

3

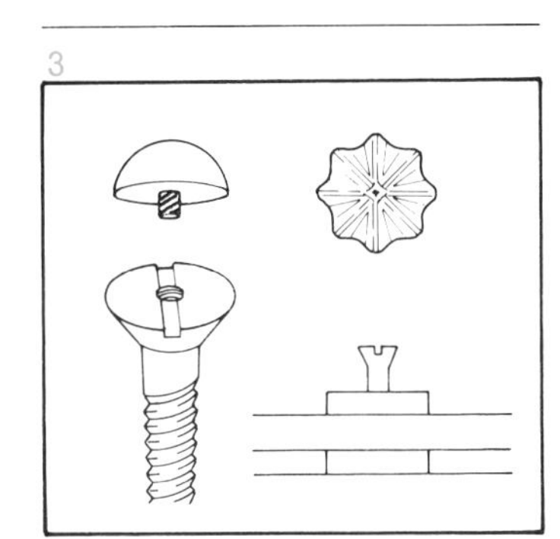

4

1 Remember mirror clips need spacing washers

2 A surface must be flat if double sided pads are to grip

3 Mirror screws can have domes, or decorative caps in a choice of colours

4 Check with your glazier whether your mirror is suitable for fixing with adhesive

Guide to double glazing

While double glazing may not be top of the pops as far as money-saving on fuel bills is concerned, you can save something like 10% by adding a second pane of glass. And there are other bonuses. Those unpleasant down-draughts which normally keep folk away from the window areas in winter are eliminated. And condensation will be drastically reduced if not prevented on window glass.

There is also a slight reduction in noise through glass, but for effective noise insulation a different spacing is required. See the section on noise.

There are a number of ways you can add a second pane of glass.

Sealed units. These come as units from the factory with two sheets of glass spaced apart and the gap between hermetically sealed. Each sealed unit replaces a single pane of glass, and most modern frames have rebates deep enough to take a sealed unit.

Secondary sash. Here a separate frame is secured to the existing window. It should be openable for cleaning and for removing any slight condensation between panes.

Coupled windows. Normally added as replacement windows which open normally but which can be separated for cleaning.

Aluminium or uPVC are the most common materials for frame construction, and in recent years big improvements have been made in the metal frames to reduce the risk of thermal transmittence through the aluminium.

1 There are four basic types of double glazing. This is the factory-sealed unit which is tailor-made for given frame sizes.
2 Secondary sash where a second pane has its own frame
3 Coupled windows. Frames are hinged together and can be separated for cleaning
4 Do-it-yourself, varying from a simple plastic channelling to sealed units

1

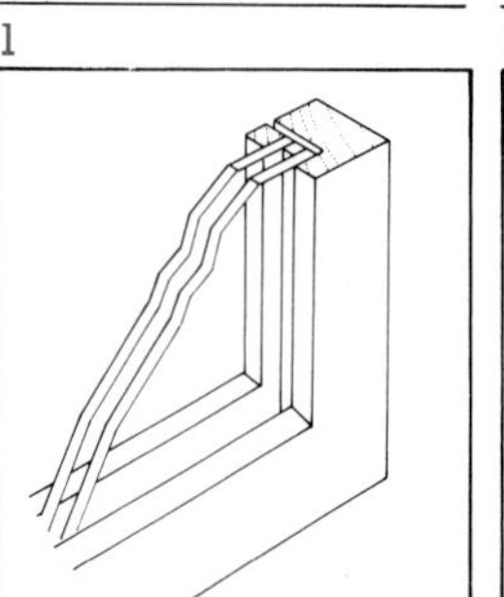

2

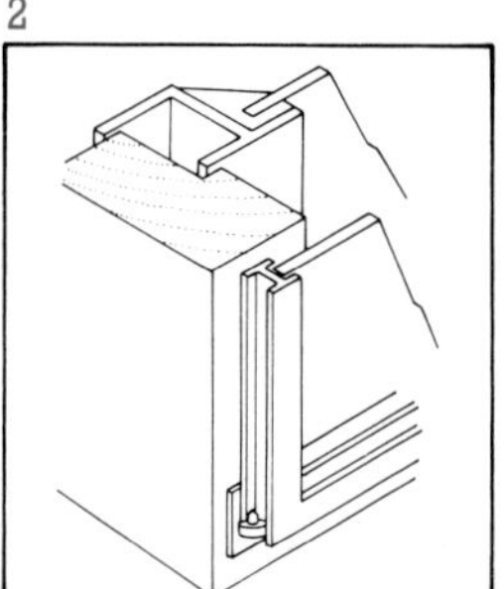

3

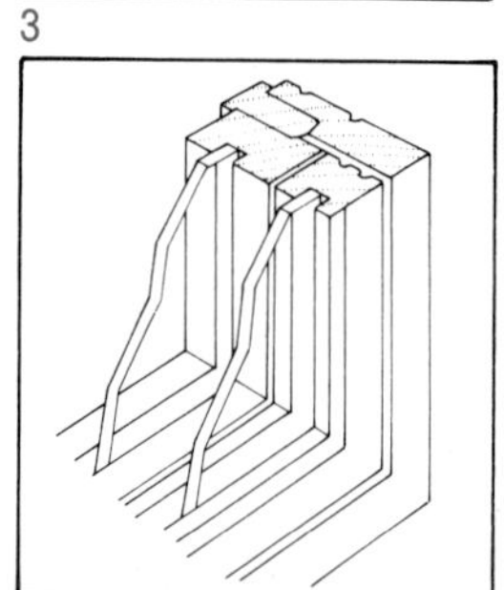

4

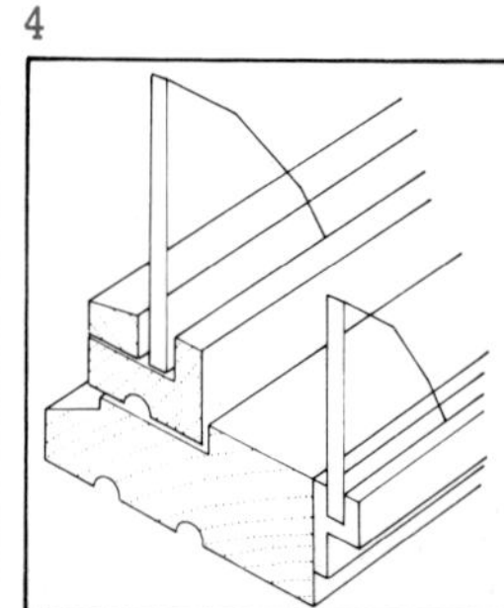

1

2

3

4

5

1 Sliding patio doors
2 Casement window
3 Upvc horizontal sliding window
4 Georgian style windows by Therm-a-stor
5 Maintaining character with windows

Guide to double glazing

A d-i-y double glazing system makes this method of insulating most cost-effective. They are designed for easy ordering, and packs can be purchased to do as little as one window at a time.

6

7

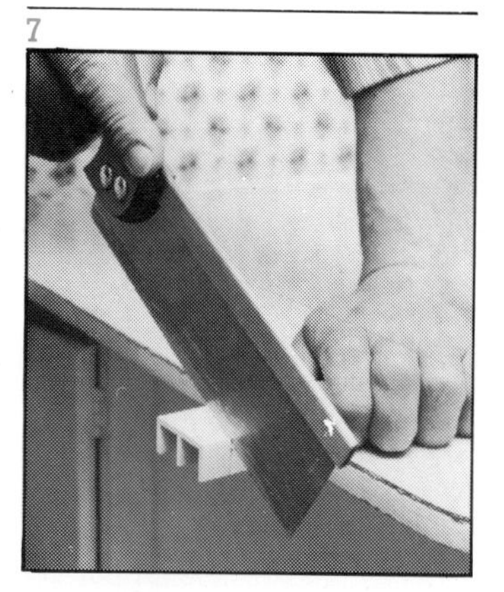

10

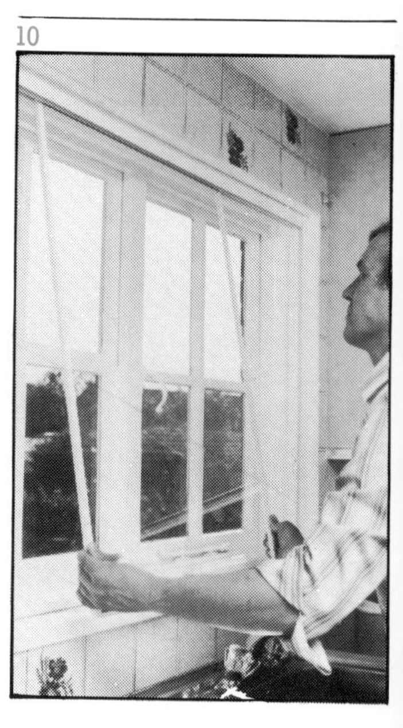

8

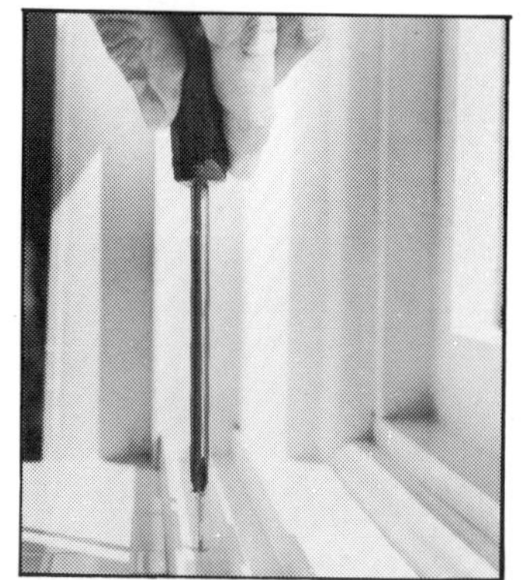

9

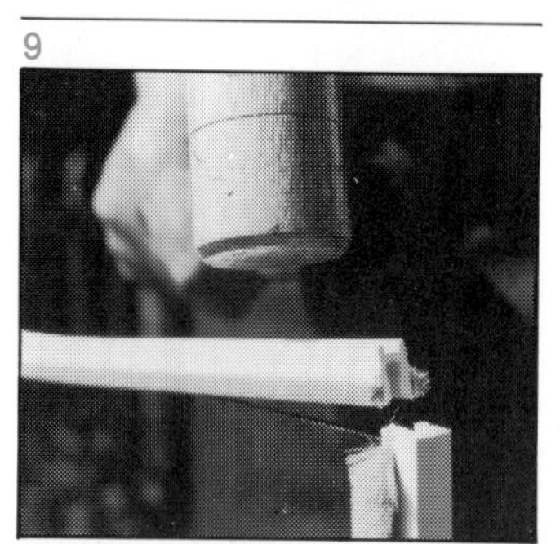

6 Windows are measured for height and width, and kits are bought to the nearest larger size
7 Tracks and frames are easily cut with a tenon saw
8 Then they are fixed in place with screws
9 You buy 4mm glass or plastic, cut to size by your dealer. Note the weather seal
10 Units simply slot on to the runners

A finished horizontal sliding system is seen opposite

11

12

13

14

11 Attractive patio door
12 Bring the garden indoors
13 Units easily opened for ventilation
14 Georgian style windows

Cutting and shaping glass

1 These are the tools you will need
A Oporto type pliers and cutter
B Or traditional wheel cutter
C This cutter will tackle circles
D Pliers for 'nibbling'
E Turps as a lubricant
F Accurate rule
G Straight-edge
H Chinagraph pencil

Cutting your own glass is perhaps not the simplest of d-i-y jobs, but with a little practice good results can be achieved. Wherever possible use new glass, or at least glass which has been well cleaned. And choose a good wheel cutter. The diamond cutter is best left for professional use as so much depends upon finding the right facet of the diamond.

The pliers-type glass cutter is a useful tool, as it can be used to both score the glass, and to grip it and apply just the right pressure along the score line to snap it clean.

Apart from the cutter, you need a straight-edge, pliers, rule, chinagraph pencil and a little turps substitute to lubricate the score.

1

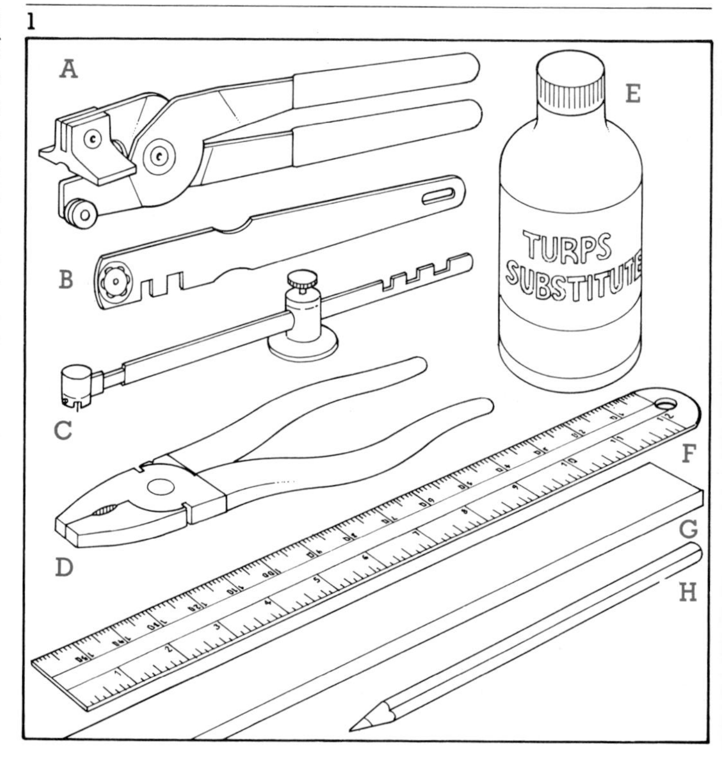

A good tip is to use a large sheet of clean newspaper as a guide under your glass. The rules on the page will be accurate, and the lines of type will give true right angles.

The illustrations show the order of events for making a cut.

1

2

3

4

5

6

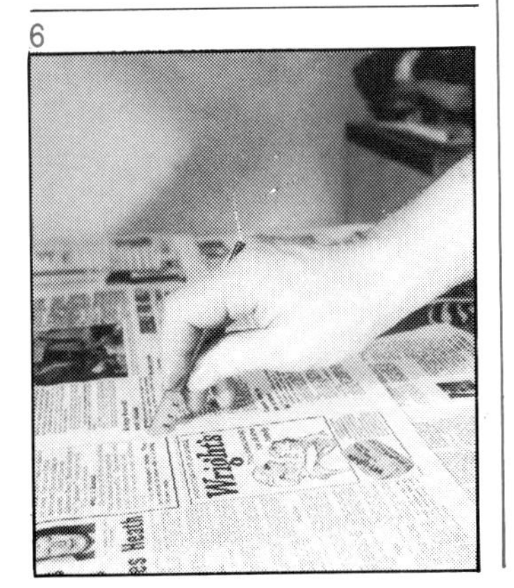

1 You need a flat surface, a soft cloth over it, then covered with newspaper
2 The rules on the paper will provide accurate angles
3 Measure twice and cut once! Mistakes can be expensive
4 Align your straight-edge as a guide for the cutter
5 The wheel should just whisper over the glass
6 Hold it between finger and thumb like this and draw it towards you

Cutting and shaping glass

7 Position over a pencil or matchsticks and press lightly either side. Or use the Oporto type pliers

8 Fine slivers can be levered away with the cutter notches

Or use pliers protected with a piece of chamois leather

10 To cut a hole for a fan, use a beam compass cutter

11 Score a second circle on the same side of the glass, B, and cut segments as at C and tap out pieces, working from uncut side

12 Make cuts D and tap out. Then make cuts F and G and tap out the remainder

7

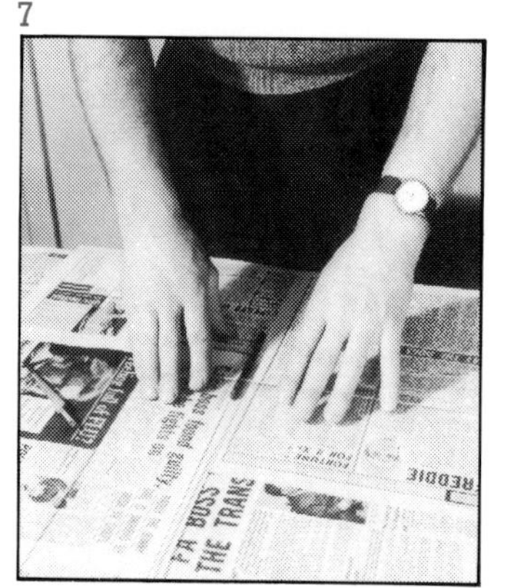

8

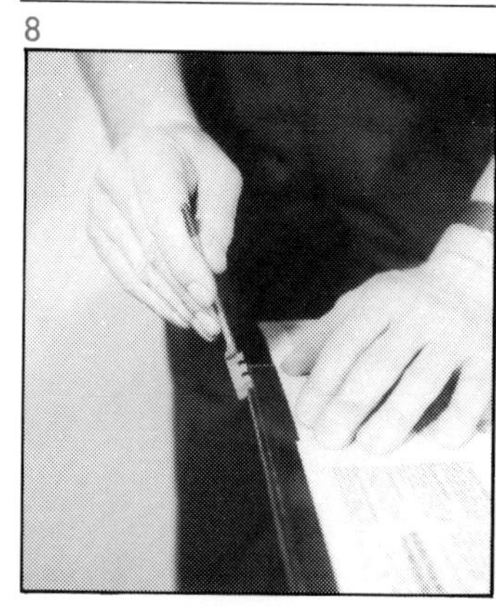

9

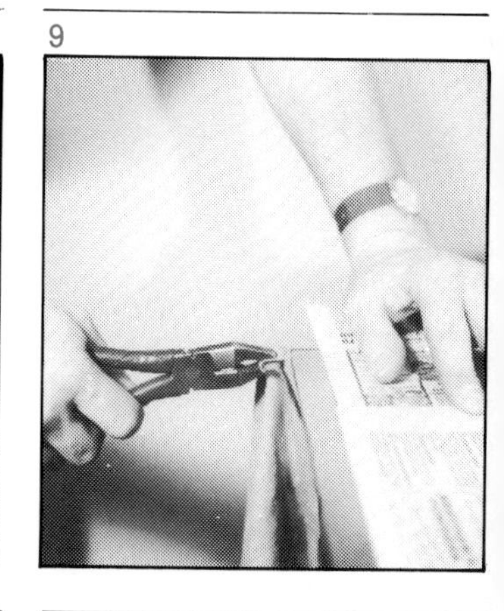

10

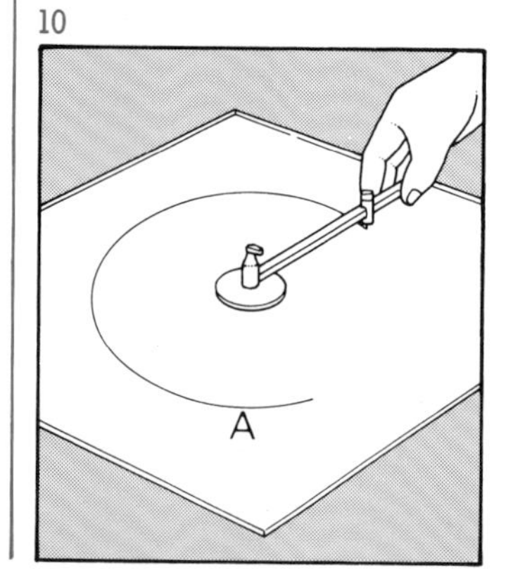

11

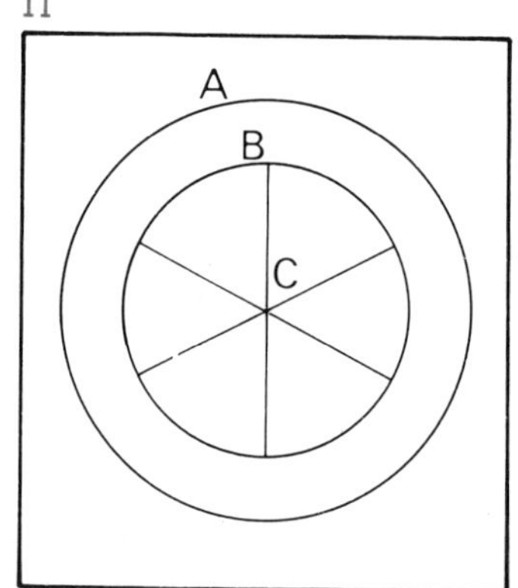

12

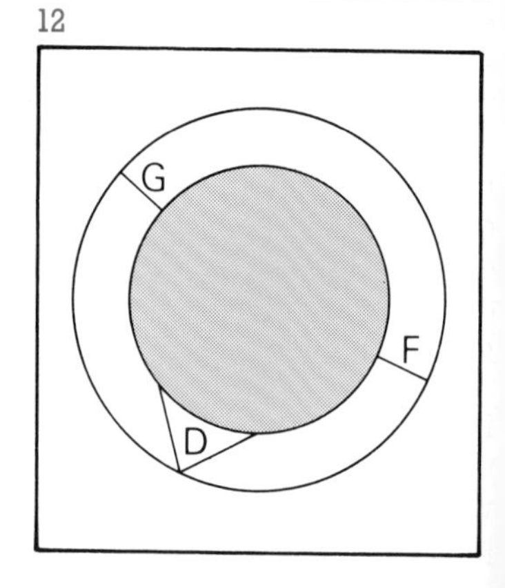

MAKING HOLES

Cutting a hole, as needed to install a fan in a window, needs a little care. Draw a circle of the required diameter on a sheet of paper and place it under the glass. Score slowly around the circle, pushing the cutter forward. Turn the glass over and press gently around the cut. This will start the cut running. Now cut a second circle on the same side as the first, then mark off in segments as shown in the drawing. Stand the glass on edge and tap out the segments with the end of your cutter working from the uncut side of the glass. Make cuts as at D and tap out gently. Make cuts F and G and tap out the remaining pieces.

You can cut a circle freehand by laying a pattern under the glass and using a wheel cutter. In this case push the cutter forward.

Mirrors for all occasions

continued from page 23

paper needs to be stripped and the surface underneath treated and prepared properly before mirror tiling can proceed.

Mirror tiles can be very useful in bathrooms, especially for covering small, limited areas. It is also safer to place a mirror on the wall behind the bathroom cabinet, than to have a mirror-fronted cabinet – with the possible risk of causing damage through falling toiletries.

It only takes a little imagination and forethought to conjure up some magical ideas with mirrors. And if there's one place in the home where some early morning magic is often needed, it's in the kitchen at breakfast time. This is an area where space is always at a premium, with the whole family jostling for room. A cleverly sited mirror, perhaps on the wall behind one of the working top areas being used as a breakfast bar, or simply on the wall behind the kitchen table, can creat an illusion of more space, with room to breathe.

Mirrors reflect not only images but also light. Poorly lit rooms can be considerably brightened by the adept placing of mirrors in sympathy with light-admitting windows or apertures. And mirrors used at the top of staircases and in dark entrance halls can greatly help to disperse light over a large surrounding area. This is a safety factor of no small significance to the elderly or poorly-sighted people.

Again, a mirror placed at a strategic point close to the front door of an entrance hall will not only bounce light back into the hall whenever the door is opened, but will also enable quicker identification of callers to the house – an important point in these days of need for security. Small, dark bedrooms, can also benefit from additional mirrors which catch and distibute light more generously.

For example, an oval mirror in an ornate frame will suit an elegant decor, whereas an oblong one may be more appropriate for a room in modern styling. A mirror framed in a decorative wooden moulding and mounted on the type of flush hardboard door so commonly found in post-war houses, can be strikingly effective.

Then there are the many artistic ways in which bronze or grey tinted mirrors may be used – perhaps to instil a feeling of intimacy in a softly coloured room, or possibly to tone down unwanted glare in a reverse situation.

Glass engraving

1 The complete engraving kit
2 Both sharp lines and fine shading can be achieved
3 Examples of work produced using the engraving kit

If you've ever admired the work of a fine glass engraver, it is encouraging to know that simple tools are available now in kit form which will enable you to take up this fascinating hobby. And you can produce attractive results in a very short space of time.

You use Swiss made high quality diamond and diamond dust tools used like fine pencils. And on soft glass they enable you to do both line and fine shading. A very wide range of patterns to copy are available, some supplied in the kits, and others available to order. The design is merely taped to the underside of the glass, then you follow the lines with the appropriate tool.

Full details of the kits are available, as are details of one day instructional courses where each person has individual attention.

1

2

3

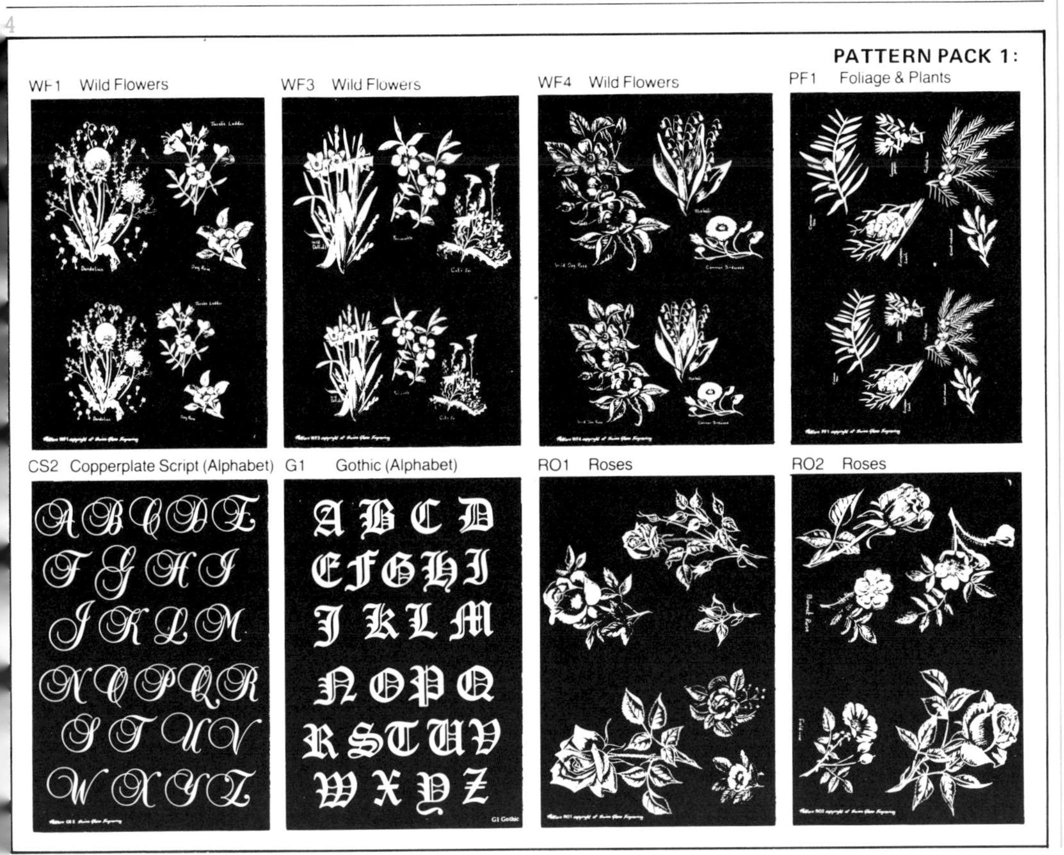

4 A very wide range of patterns is offered, including lettering

For your reference

CHOOSING GLASS

LOCATION	TYPE	MM THICKNESS
Greenhouses	Horticultural	3
Picture frames	★ sheet (or diffuse reflection)	2 or 3
Windows up to 6sq.ft.	float	3
Windows up to 12sq.ft.	float	4
Windows up to 6 sq.ft.	float	5
Windows up to 25sq.ft.	float	6
Windows up to 50sq.ft and patio doors	float	10
Area at risk	toughened or laminated	5
Glass doors (internal)	toughened	
Shelves & table tops	float	5-10
Skylights	wired	6

Thicknesses are only a guide. In exposed positions, wind pressure must be allowed for when thickness is chosen. Also ask the glazier re shelf thickness for given lengths.

★ Very little sheet glass is available now. Float glass is the most common. Plate glass is expensive, so float usually takes its place in domestic situations.

WOOD AND WOODFINISHING

Contents

Simply beautiful

There is something rather special about wood. It is a pleasant material to work with; it can be formed into the most attractive shapes, and it is good to feel and to handle. No man-made materials can match it, and it is one of the few resources which, by careful husbanding, can be replaced as it is used – certainly as far as softwoods are concerned.

Yet timber presents something of an enigma, for while it has suffered a bad reputation in the past for rotting, history shows that properly protected it will last for a thousand years. Sadly the timber trade have done little in the past to protect its reputation, and in many cases they have allowed materials like aluminium and plastics gain an unfair advantage.

Despite all this, it is encouraging to note that there is a revival of interest in the use of natural timber, and in the use of wood finishes which enhance the product rather than hiding it under a coating of paint. Doors covered with hardboard 15 years ago are being cleaned off and natural finished. Banisters are stripped and sealed, while old, sound furniture is being given a new lease of life.

Natural beauty

There are so many types of timber that it is impossible to give a comprehensive list of their properties, colours and working qualities in such a small space. However, a short list of some of the more popular timbers on the market will prove useful.

Ash: Two main countries supply this timber, the U.K. and America. The British variety is mainly white with a fairly coarse texture, but the heartwood is often light brown. The timber is very tough and springy, it can be finished nicely and is resistant to splitting. The American version is similar and it is normally white with a coarser grain.

Beech: A straight-grain, fine-texture timber with easily recognisable dark flecks. It is light reddish brown, sometimes darker brown, in colour and moderately hard. It is an easy timber to work with and has a wide range of interior applications from turning to furniture. Not recommended for outdoor use.

Boxwood: There cannot be a woodworker who hasn't heard of boxwood tool handles. This is a dense, heavy timber with a very fine grain which makes it extremely useful for turning work. Available only in fairly small pieces, as the wood comes from a bush rather than a tree.

Cedar: Probably the best known variety of cedar is western red cedar which comes from British Columbia. But there are many varieties, including some from America and Canada. Colours range from pink to reddish brown and it has a distinctive smell which many woodworkers find attractive. The grain is fine and the timber works well.

Chestnut: A hard, heavy and tough timber which, in many ways resembles oak – but its weight is somewhat less. It is useful for outside work but it does contain chemicals which, in the presence of moisture, will corrode metal fittings. Best uses are for internal work.

Ebony: While ebony is used as a term for something black, in fact the timber is a medium to dark brown colour with black stripes; it may also be mottled with grey or brown marks. The timber is dense and a good material for turning and carving.

Elm: Dutch elm disease must be known by everyone in the country as this was the most disastrous event for the elm population. That which survived has a dull brown/reddish colour – but it is extremely hard to find. More freely available is its northern counterpart which is straight-grained and has a white sapwood with a green streak in it. The timber is tough and works well; it accepts both stain and polish well.

Iroko: Yellow-brown timber with a coarse grain. Extremely durable with a vast range of applications from furniture to outside structural work.

Larch: Reddish-brown fairly hard timber which is tough and durable. Grain is straight and fairly free from knots. It has a natural resistance to rot and is mainly used for outdoor applications such as fencing and boating

1 European elm; a yellow-brown timber which can be difficult to work but stains and polishes well
2 Douglas fir; easy to work, this red-brown timber finishes well
3 European whitewood; a non-resinous white to pale straw colour timber
4 Central American Mahogany; a very pleasant timber to work, this red-brown timber is ideal for cabinet work
5 European birch; easy to work, this light-coloured timber stains and polishes well
6 Parana pine; light brown colour, easy to use but likely to warp. Takes good finish
7 Western red cedar; easy to use, this timber finishes well but grain does rise
8 Teak; greasy to the touch, teak can be moderately difficult to work but polishes well

Natural beauty

9, 10, 11 Three patterns of hardwood flooring. First is a mosaic, then a herringbone and finally strip
12 Varnished floor with a loose mat sets the style while the chest, made in a similar material, is subtly altered with coloured stain

9

10

11

12

Lignum vitae: Extremely fine texture and grain combined with great density (80% heavier than oak) makes this an ideal timber for turning. Colour is a distinctive green-black and the high resin content gives the wood a self-lubricating quality which makes it useful for certain shaft bearing applications.

Mahogany: Probably Honduras mahogany is the one which evokes the most admiration, but it is now scarce. But there are many varieties from America, Australia and Africa with colours varying from almost colourless to a rich deep brown. Grain is fine, but irregularities which give attractive figuring are common. It works extremely well and is used for high-class work especially in the boatbuilding industry.

Oak: British oak is, of course, the best known, but there are hundreds of varieties coming from as far as Tasmania, Japan and America. Hard, tough and durable, oak is a light brown timber which is ideal for heavy structural work as well as decorative interior applications. It polishes well but the porous nature of the grain is such that paint is soon pushed off.

Parana pine: A light golden-brown softwood that works well giving a fine finish. Being almost knot-free, it is much liked for joinery and furniture work. Difficult to season, it tends to split and warp.

Ramin: Really straight grain and an almost white appearance makes this an ideal timber for mouldings. Accepts stains well but it is rather brittle and care has to be exercised when pinning mouldings made from it in place.

Rosewood: A dark, beautiful timber which is hard, heavy and ideal for decorative work. Although India is the prime rosewood country, supplies also come from countries such as Brazil and Madagascar.

Sapele: A difficult timber to work, it is strong and durable with a white to pale yellow colour. Freshly sawn timber has a smell similar to that of cedar. Used for furniture and quality joinery; also cut into veneers.

Teak: Originally from Burma, the true teak is a golden-yellow to golden-brown colour. It has a greasy texture while the grain is straight. Slow to season, it is durable and warp-resistant. It is difficult to work and is used in ship-building applications, for furniture and as a veneer.

Walnut: Wide range of colours from pale yellow to nearly black depending on country of origin. British walnut is light brown to black and is a strong, tough durable material with a fine grain and attractive figuring. Its American counterpart goes from yellow to chocolate brown. Uses include decorative furniture, turning and carving.

Yew: An extremely durable softwood with white to red colours. Fine grain which makes it ideal for turnery and carving work. Timber is very springy which is why it was used to make battle-winning longbows in Olde Englande.

Reformed timber

Man-made boards is perhaps a misleading term because they are in fact made from timber. But on some only close examination will reveal the timber because it has been cut up, mashed, squashed and treated in a number of ways before being impregnated with adhesives and pressed into boards under great pressure.

Of course, there are boards which are much more recognisable as timber – plywoods and blockboards being examples. So let us look at each type in turn.

Hardboard

Wood fibres, separated into tiny portions, are used to make hardboards. The finished product is hard and smooth on one side while the other side is comparatively rough. There are many types of hardboard and they are used for panelling, floors, (as a surfacer or underlay), for decoration (enamelled, embossed and painted types), insulation (both sound and heat) and utility applications such as pegboard for displays and storage.

Before using, most hardboards need conditioning. This merely means allowing the board to adjust itself to the moisture content of the place in which it is to be used. For ordinary hardboard in living rooms, this means standing them, exposed to the air, for 48 hours.

At the other end of the scale, for damp places such as bathrooms, between one and two pints of water are brushed on the back of each 2440×1220mm sheet and the sheet is allowed to stand in the room for 48 hours. Tempered boards need 72 hours for conditioning.

Hardboard has one other important property; it can be bent easily to shallow curves by hand and, for more acute bends, it can be soaked first and then bent.

Sizes available in standard hardboard usually start at 1220×1830mm, going up to 1220×3660, with door sizes also offered. Thicknesses start at 3mm going up to 6mm.

Grades include standard, tempered (with greater strength and water resistance), enamelled (on one side for kitchen and bathroom applications) with plain or tile patterns, moulded (in a wide variety of designs such as reeded and woodgrains for decorative use), and perforated (for use with special clips and accessories for holding tools and other items).

Insulating board is a lightweight type of hardboard and, as such, falls into this category. It is usually between 12 and 25mm thick and in a wide range of sizes suitable for use as ceiling tiles or wall boards. Two types are made, one which is primarily intended for sound absorption, (not insulation), the other for heat insulation. Those intended for exposed use, such as ceiling tiles, are suitably decorated on one side.

Plywood

Considerable strength is one of the main properties of plywood, the reason being that each of its thin plies lies at right-angles to the plies on either side of it. It also has resistance to warping and this is because there is always an odd number of plies to balance any tendency to bend. By choosing grades, other properties can be selected; manufactur-

1

2

3

4

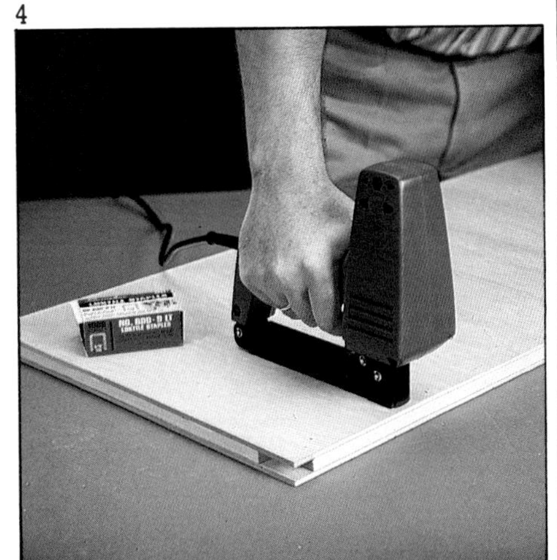

1 Attractive home for plants made from a combination of boards
2 Wallboard provides a matching background for the shelving units made from kits
4 It would take more than a casual observer to note that these shelves are not made of real timber
3 Extremely strong sandwich construction of thin plywood and battens, glued and stapled together

Reformed timber

ing treatment can make plywood resistant to the weather and extremes of heat; various timbers used for the final layers of veneer can be chosen for excellent decorative effects.

Thicknesses range from 3mm to 25mm, roughly in steps of 3mm. The thinnest types can be bent easily to small radii. Board sizes are normally 1200×2400mm although almost any size can be obtained, being cut to your requirements by the timber merchant.

Each type of plywood comes in a number of grades. For example, boards with Finnish birch as the final ply are suitable for finishing to a high standard. Face qualities are defined by letters. B quality is the best, ready for staining, polishing or high quality painting. BB is the standard painting grade or for other surface finish. WG quality is for textured finishes or as the opposite face to a better quality – one side may be finished to B standard while the opposite face could be WG. E grade is for pine finished boards, and these are suitable for clear finishes to expose the wood grain.

In addition there is a

BOARD EDGINGS

Lipped before veneering
Lipped after veneering
Tongue and groove
Increasing thickness
Edge veneered
Hardwood lipped
Extruded tongued plastic
Extruded aluminium, edge of top laminate trapped

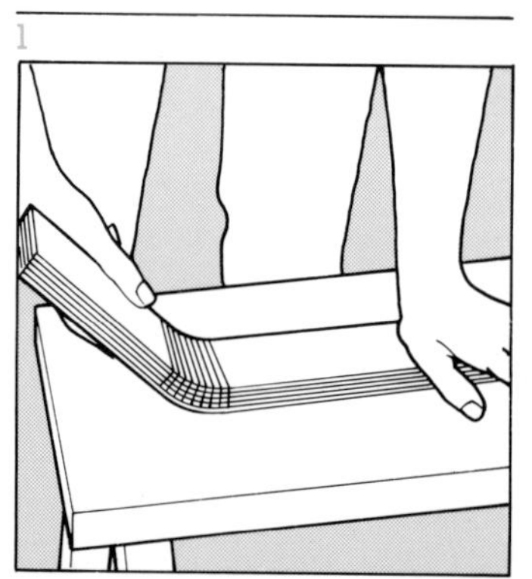

1 A series of fine cuts makes bending simple. They must stop at the final laminate

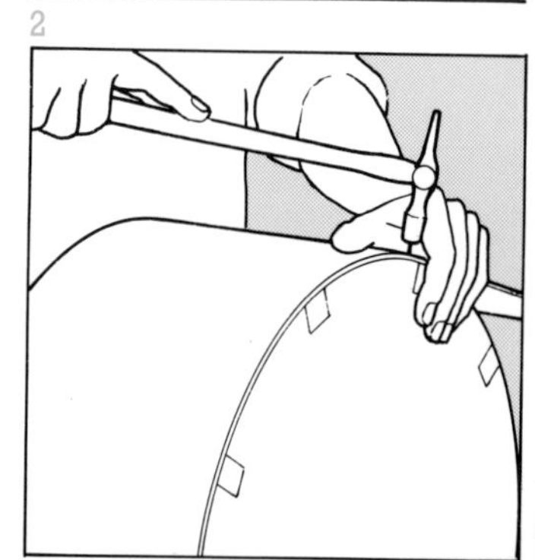

2 Damping standard hardboard simplifies bending. It doesn't work with oil-tempered boards which are water-resistant

number of special finishes; surfaces such as phenolic or plastic films gives protection against various liquids; GRP or metal surfaces give even tougher protection; some boards can be painted, varnished or treated against the spread of fire. Preservative treated boards can be used in permanently damp conditions.

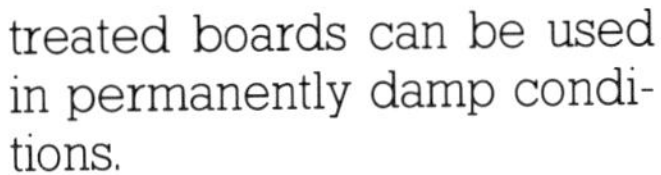

The plies are joined together with adhesive and, depending on the quality of this adhesive, the board is given a rating. INT is the standard interior quality,

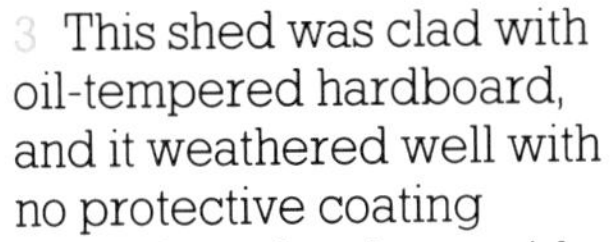

3 This shed was clad with oil-tempered hardboard, and it weathered well with no protective coating

4 Pegboard makes an ideal backing to which tools can be clipped

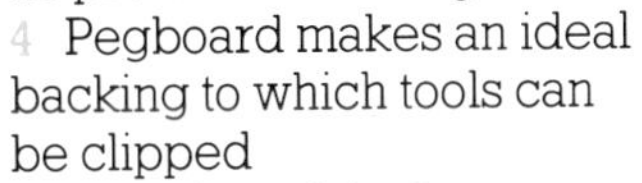

5 Just a few of the large range of chipboards available

6 Boards of laminate construction have immense strength

suitable only for interior work. MR indicates moisture resistant which means that the board will withstand full exposure to weather for limited periods, will withstand cold water for a reasonable length of time, hot water for a much shorter time and will fail under boiling water. BR stands for boil resistant, so that the board will withstand weather and boiling water for prolonged periods. It will also withstand cold water for extended time. WPB is the top grade and this has strong resistance to cold and boiling water, dry and steam heat and weather.

Blockboard

Much cheaper than plywood, blockboard and its related laminboard and battenboard are made from two outside veneers bonded to a solid softwood core. There are carefully laid-down definitions for each type of board, but briefly the differences are as follows.

Blockboard is made of a collection of sizes ot timber to make up the core. Battenboard is smaller, but the core material is of higher quality and of certain, larger, sizes. Laminboard is made of carefully sized laminations of redwood or birch for the core.

Standard board size is 1220×2440mm, while a variety of thicknesses and finishes are offered.

Chipboard

Probably the most used type of board, it is available in a vast range of finishes, sizes and thicknesses. Grades include fire-retardent and water-resistant, grades with various finishes, and melamine and timber veneer faced materials are now so common both on furniture and in the high street shops that they need no introduction.

The board is made from particles of wood, coated with resin adhesive and bonded into sheet under heat and pressure. The resin is very tough on sharp-edged tools, so tungsten-carbide tipped saws and planer blades are often used. While it is described as a homogenous material, usually the outer faces are extremely hard and tough while the core is much more friable. For this reason, fixings into chipboard are more difficult than conventional materials – but see our section on joining up.

The two main categories are standard and flooring grade. The latter is much heavier and more dense and therefore capable of carrying heavy loads.

Three versions of flooring grade are made; square-edge, tongue and groove with an integral tongue, and grooved ready to accept a loose tongue.

Thicknesses go from as little as 4.5mm up to about 25mm (for retail sale) while the standard board size is the usual 1220×2440mm. Few additional sizes are offered except as offcuts, although flooring grades come in half-sheet sizes.

When surfaced chipboard is considered, then the choice becomes almost limitless. Obvious examples are veneered and melamine faced, but plastic laminates, sprayed-on finishes, photographed-on wood grain finishes even metallised surfaces are on offer. Sizes and thicknesses of these specials are also special, so contact manufacturers or suppliers before making plans.

1

2

3

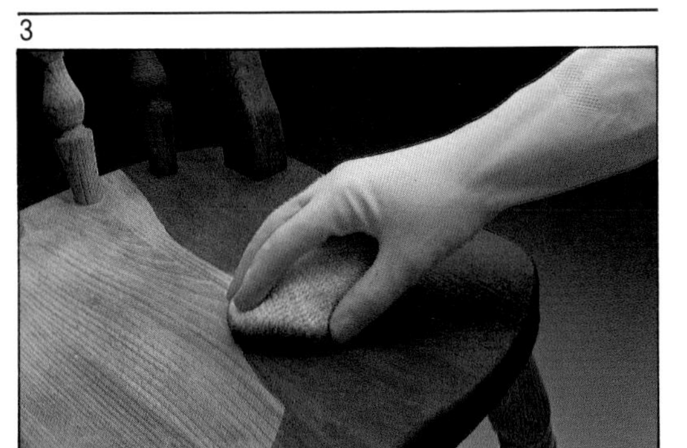

4

5

1 Applying a coat of interior varnish on to a clear varnish base coat
2 Colour preservative protects and improves the appearance of this wooden wheelbarrow
3 A cloth pad being used to apply wood dye to a wooden chair
4 Varnishing a handrail after removing an accumulation of years of painting
5 Cleaned and varnished, the stairs present a fashionable pine finish

Finishing touches

Stains

The natural beauty of wood, especially hardwood, needs to be protected in a manner that allows the grain pattern to show through. The preparation of timber ready for the finishing process is completed by sanding the surface to very smooth finish. A filler of the appropriate colour, thinned to a creamy consistency with white spirit is used to fill the open grain of timbers such as oak. A flour-grade paper is used to produce a perfect surface.

Before repolishing starts remove old grease and wax

The colour of the wood is often enhanced by applying a liquid stain. This can be either water or spirit based. Always try out the stain on a piece of waste wood before use. Water based stains can be diluted with water and it is best to do this and build up with several applications rather than try to get the right shade in one coat. Always ensure that the stain and the finish are compatible. This is best done by using the same manufacturer's materials and following his recommendations.

These stains are not to be confused with the polyurethane stains which do not stain the wood, but are in fact a coloured polyurethane varnish. Coloured varnishes cannot be lightened by thinning, but they are darkened by extra coatings.

Points to watch when staining timber are to avoid drips running on to end grain as it will dry out darker. Take care with overlaps and have a cloth ready to wipe off excess as it will dry darker. Finish end grain using a half-diluted stain so that it will dry the same colour as the rest of the wood. A wet rag can be used to even out uneven water stain. Leave the timber in the room to dry out thoroughly before starting the next stage.

Finishes

Polyurethane varnishes are probably the most popular finishes, because they dry quickly and produce a hard, heat resistant surface. They are brush applied and are made in gloss, satin or matt finish.

A first coat or primer is diluted with white spirit according to manufacturer's instructions. A very fine abrasive paper can be used between coats. Always use a soft, good quality brush flowing the coating and using the minimum amount of brushing out to level it.

Two-part cold cure lacquers are resin based materials which have a hardener added to them just before use. They produce a hard, non-yellowing finish with a high gloss. They are ideal for table tops and surfaces which have to withstand heat and solvents.

Wax polishes are usually used over other finishes to improve the shine, but they can be applied to new wood. Polishing bare wood does take some time and it is best to seal the wood with a coat of French polish or diluted polyurethane, this especially applies to oak and similar open grain timbers.

You can make your own wax polish by dissolving beeswax in genuine turpentine (not white spirit) until it becomes like thick cream. The process can be hastened by warming the container in hot water, but do not heat it over an open flame as the fumes given off would be inflammable.

The wax can be applied by brush or cloth. It then has to be rubbed vigorously with a cloth or a fairly stiff brush. Over the course of time and many applications a deep, antique type of shine will result.

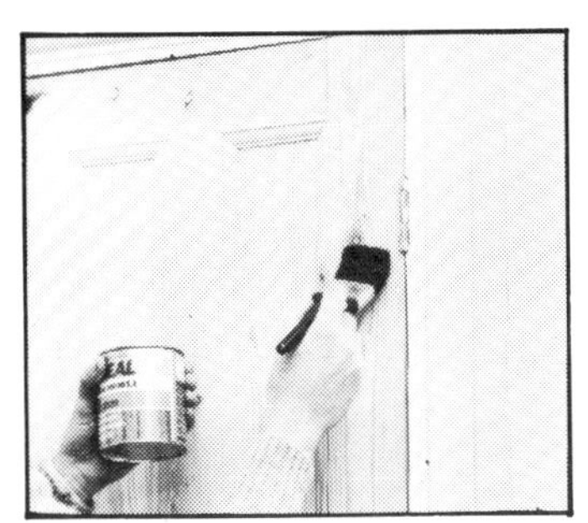

Applying varnish to a stripped pine door

French polish dries quickly and a fair degree of skill is required to obtain a good finish. The normal type has a very light yellowy-brown tint, but you can get white polish and this is almost colourless. Application is by a cloth pad made by wrapping a piece of lint-free cotton rag around a wad of cottonwool. The wadding is saturated with the polish and is then pressed gently on a spare piece of wood to remove the excess polish and flatten the face of the pad.

A few spots of linseed oil are put on the face of the rag to act as a lubricant. The pad is then rubbed over the surface of the wood in small circular movements, gradually increasing the pressure to force out the polish. When the polish is used up allow time for the surface to dry before restarting the application. With patience, a high gloss finish can be achieved.

Burma teak is a naturally oily timber which is very durable. Like most other timbers exposure to sunlight and weather, when outdoors, causes it to lose its good looks. Proprietary teak oil can be applied using a soft rag. A lot of rubbing with the grain will be needed to produce a low gloss finish. Although called teak oils, they are suitable for other woods and in particular iroko which looks like teak, but it has not got the naturally greasy surface. Both these timbers being very durable are used for outdoor furniture and will, therefore, benefit from oiling to maintain or restore their natural appearance.

A clear finish for exterior woodwork, particularly hardwoods, is yacht varnish. Specially formulated for severe exposure, this varnish is more elastic than polyurethane and will cope with natural movements in the timber. Polyurethane is also a good varnish as it gives a highly glossy and hardwearing finish. It is ideal for the more sheltered places.

Water repellent wood preservatives are made both clear and in various colours. They can be used for window frames, sheds and railings instead of paint and they have the advantage that the grain in pattern of hardwoods is not covered up as it is with paint. They can, however, only be applied to bare wood. Painted timber must be stripped.

Timber, everywhere...

Modern materials may be strong, durable, cheap, easy to produce and reliable. But they can also be boring! Timber, on the other hand, is beautiful and has been the foundation of craftsmanship in bygone years and provides hours of pleasure for both makers and users in today's monotonous world. Here are some examples of the beautiful use of timber.

1 An attractive ornament for the garden – but the timber must be well protected with preservative
2 Alicante sapele door blends well with the stone surround
3 What better way to store wine but in a glass-top hardwood table?
4 Outdoor again, and here whitewood is used to good effect
5 Stairs, rails and spindles nicely matched

1

2

3

4

5

6

7

8

9

10

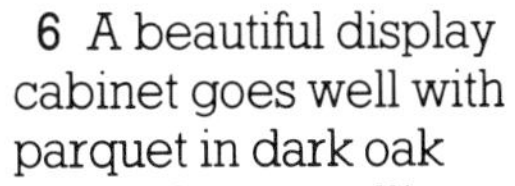

6 A beautiful display cabinet goes well with parquet in dark oak
7 Timber panelling on the ceiling and strip flooring complement the painted door and bookcase
8 Oak kitchen units together with matching table and seat unit make a pleasant ensemble
9 Knotty pine abounds in this modern bathroom
10 Olde worlde living room with timber beams, open fire and modern accessories for comfortable living

Timber, everywhere...

11 Oak, traditional in the kitchen, is shown off to best advantage
12 White melamine in this kitchen is set off by the knotty pine panelled ceiling
13 Light and airy entrance hall with timber surrounds and white spiral staircase
14 Lots of lovely timber in this elegant hall/living room
15 Knotty pine panelling and ceramic tiles go well together in this modern bathroom

11

12

13

14

15

16

17

18

19

20

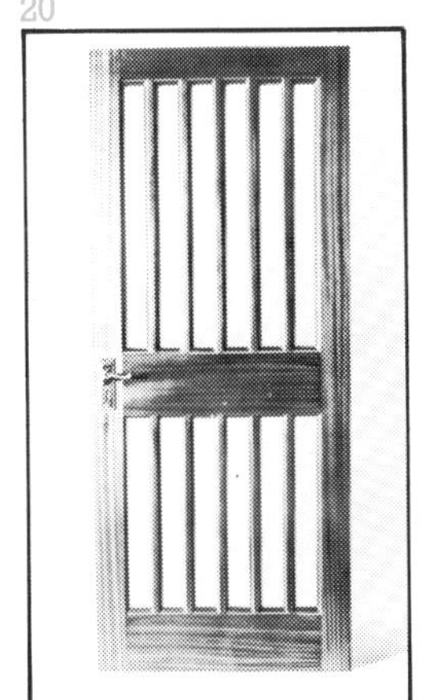

21

22

16 Left: Seville redwood door. Right: Alicante in sapele

17 **and** 18 Loose sills mean that the windows can be turned sideways and fitted with grilles (17) or used in the normal position (18)

19 Douglas fir is used for the bow front door

20 Phillipine mahogany door

21 Phillipine mahogany is also used in the construction of this door

22 Double glazed leaded lights go well with this hardwood window

Making the joins

Joining both genuine timber as well as the so-called reconstituted materials together requires a basic knowledge of their structures so that maximum strength can be achieved. Some methods can be used for both types of material, but let us take each in turn and consider ways of joining them.

Timber

Traditional joints always come to mind when making things with timber – and there is nothing wrong with this. Joints can be made to form part of the design, such as neatly made dovetail joints. However, the more complex the joint the more time-consuming it is to make, and only when maximum strength or decorative effect is needed is such a joint justified.

Where great strength is not needed, butt joints can be used, with rigidity added by nails or screws. With nails, only basic carpentry can be considered because it is very difficult to hide them; screws can be hidden in several ways. They can be countersunk and covered with wood plugs or fillers; veneers can be added if needed. Several manufacturers offer screw caps – plastic screw cups with snap-on lids. Another alternative when using Pozidriv screws is the GKN screw cap which is a press fit in the screwdriver recess.

Of course, screws can be used without glue, but greater strength is achieved with both. Glue can be used without any other fixing, for example with the old-fashioned rubbed joint. Here, the surfaces are planed to a really good fit, a thin layer of glue applied and the joint faces rubbed together, allowing the adhesive to ooze out and give a really thin glue line with remarkable strength.

Dowels are used by many do-it-yourselfers because, when fitted with the aid of suitable jigs, they are quick, accurate and extremely strong.

Dowels used to be off-cuts from a long strip of dowel rod and you can still use this method with a few precautions. First, the end of each dowel needs to have a slight chamfer (a pencil sharpener is ideal for this job) to guide it into the hole.

Very important when putting such dowels into blind holes is to provide some means of escape for the adhesive otherwise the hydraulic pressure caused by forcing a tight-fitting dowel into an accurate hole can split a piece of timber apart.

The way this is done is to make a saw-cut along the length of the dowel, but this is time-consuming. Another method is to make up a simple jig; a hole, the same size as the dowel diameter, is bored in a piece of timber and a screw driven in so that the point protrudes into the hole. The dowel (or the whole length of dowel rod) is forced through the hole and the screw point forms a vee-groove for the adhesive to escape.

However, if you buy ready-made dowels, almost certainly they will be of the serrated type. Not only do these allow the adhesive to escape up the sides, but the serrations crush as they enter the hole and, together with the glue, form an extremely good joint.

The fitting of dowels can be a problem unless you have a jig. The only excep-

1

2

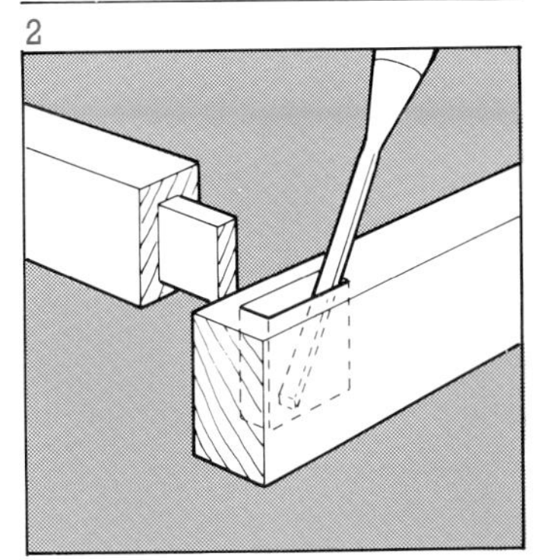

3

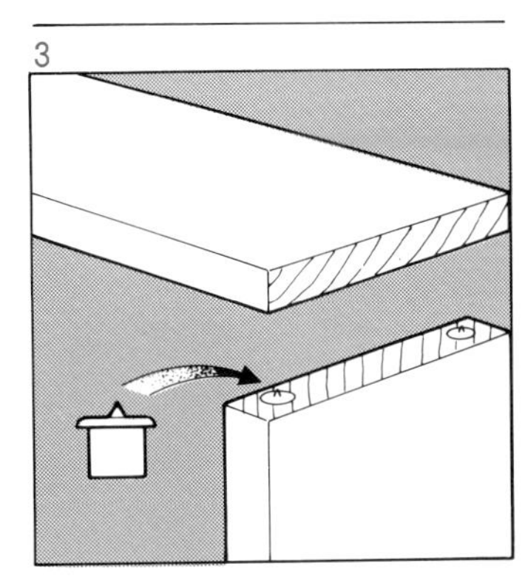

4

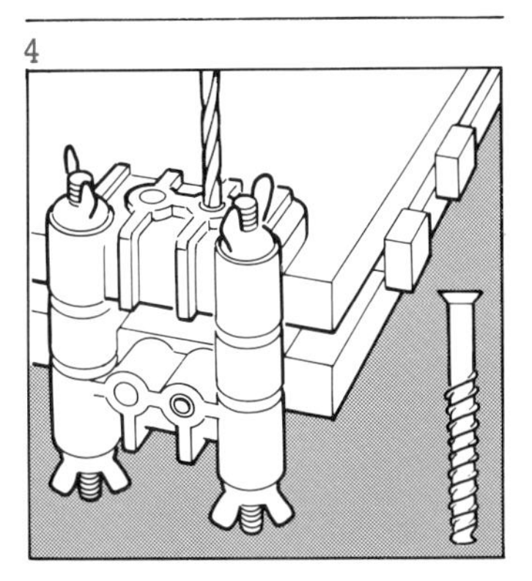

5

6

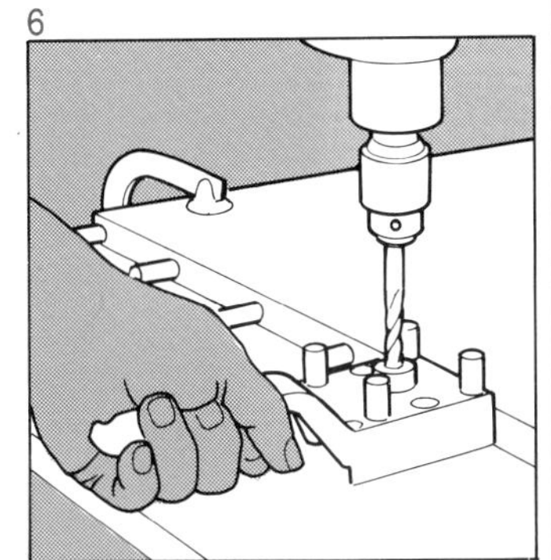

1 Traditional joints, such as this stopped dovetail, can be used only on real timber
2 Another conventional joint, the stopped mortise and tenon, used for joining rails
3 Making dowel joints using points. Holes in one half of the joint are made first
4 The jointing system for man-made boards of two thicknesses only, 15 and 16mm
5 Three sizes of dowels and a wide range of board sizes are accommodated by this jig for joining boards
6 Another jig copes even with joining down the middle of boards

Making the joins

1 Making a dowel joint using a jig. Note the depth stop on the bit
2 Assembling a dowel joint in a wooden frame
3 Building furniture using the plastic knock-down fittings
4 Using a Copydex Joint Master to make a halved joint in timber
5 The mortice and tenon joints of frames can be pulled tight using a windlass

1

2

3

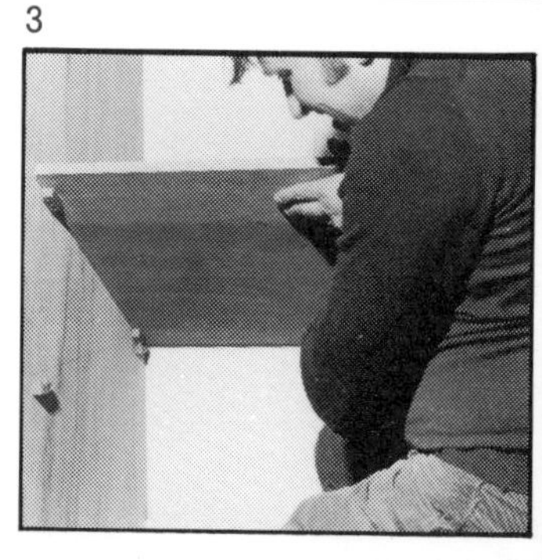

4

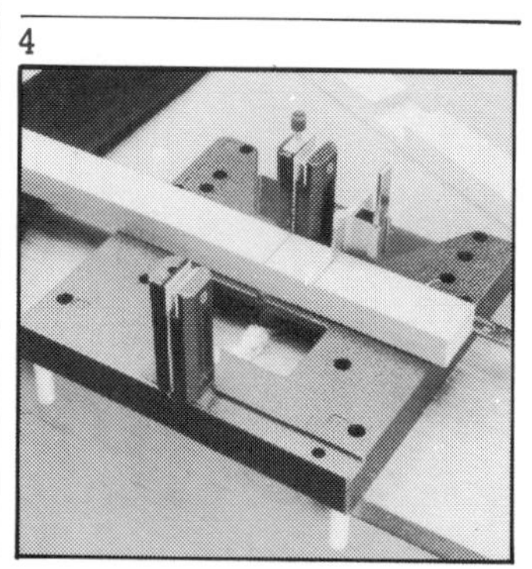

5

tion to this rule is when fitting through dowels – where the dowel is visible when the joint is complete. Here it is a case of assembling the components dry, drilling right through the two parts and finally reassembling with adhesive.

Various jigs are on the market for making dowel joints in various situations – edge to edge, edge to middle, joining battens and end-to-end butt joints. The main things to remember when buying a jig is that it will hold the two halves of a joint firmly together while you drill the holes. The exception to this rule is when working in the middle of sheet materials – in this case it is more usual to use the dowels already positioned in one half of the joint to locate the second set of holes by means of a hand-held jig.

When all other methods fail, there are devices called points. These are short metal dowels with a flange on one end to prevent them going right into the hole and a sharp point in the centre. In use, the holes in one half of the joint are drilled and the points fitted into them. The other half of the joint is offered up and lightly pressed on to the points. The marks made are locations for the matching set of holes.

Man-made materials

Really, these are reconstituted from timber waste products and are formed under intense pressure with bonding resins. These resins are extremely tough, but while the surfaces of boards so made are hard and smooth, the cores are usually fairly soft and not able to take heavy loads. Exceptions to this are plywoods, but strictly these are not reconstituted materials.

Conventional jointing is limited to tongues and grooves, often with the tongues being of the loose type and made out of solid timber. Housing joints are also fairly strong.

When using screws and glue, it is better to use special chipboard screws which give a better anchorage in the looser chipboard core. Ordinary wood screws will work, of course, but the joint isn't so strong.

A better grip is provided by drilling a hole and plugging it with a wall plug, but best of all are specially made plugs; but the size of hole you have to drill for these is rather large.

A secure anchorage on the edge of a chipboard sheet – for hinges for example – can be made by gluing and pinning a strip of timber to the edge.

Dowels can be used to very good effect with chipboard and similar materials and much the same rules apply as for real timber, mentioned above. One excellent new development in this field is the GKN metal dowel screw for which the company also offer a drilling jig. Used as instructed, these provide a really excellent joint.

Knock-down fittings

Finally, for both real timber and man-made materials, there are knock-down fittings. These are specially designed for ease of assembly and, as the name suggests, ease of dismantling. Fine if you have ideas of moving house frequently, but these have two disadvantages. They are fairly expensive and the work involved in fitting them can be more than making other types of joint.

Tools to use

1 A whetstone is ideal for grinding plane irons and chisles
2 A t.c.t. (tungsten carbide tipped) cicular saw blade; ideal for working with chipboard

There is an old saying about a bad workman blaming his tools, but the converse applies, too – no matter how good a workman you are, without good tools you cannot produce first-class work.

When selecting tools for woodworking – or for any other job for that matter – it pays in the long run to buy really good tools made by reputable manufacturers. Having bought them it makes sense to keep them in good condition by correct storage (away from damp) and by regular, efficient maintenance.

Sharp tools are mistakenly regarded as dangerous by those without sufficient knowledge of the subject – it is as well to remember that dull tools are the really dangerous ones because with them it is easy to apply incorrect techniques to overcome the disadvantages of dullness. That's when accidents happen.

A honing stone is a good investment for the woodworker, as is an electric whetstone or a double-edged grinder.

With sharp tools, working with timber becomes a pleasure, with curly timber shavings decorating the workbench after a session.

Hardwood works particularly well with hand tools and, surprisingly, softwoods need sharper tools and more care.

When it comes to working with man-made materials (or reconstituted timber products as the trade would like us to describe them) then the extremely tough resins used in their manufacture will quickly blunt ordinary tools, so t.c.t. tools are often used.

1

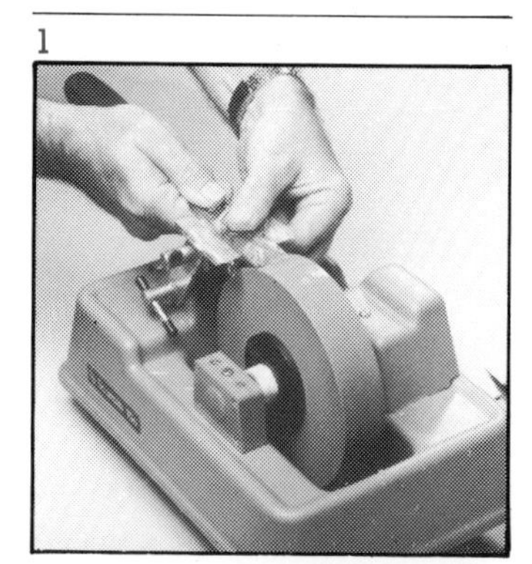

2

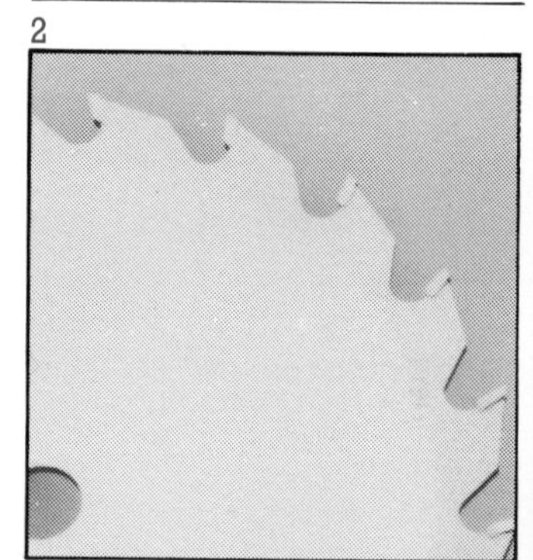

Enemies at bay!

1

2

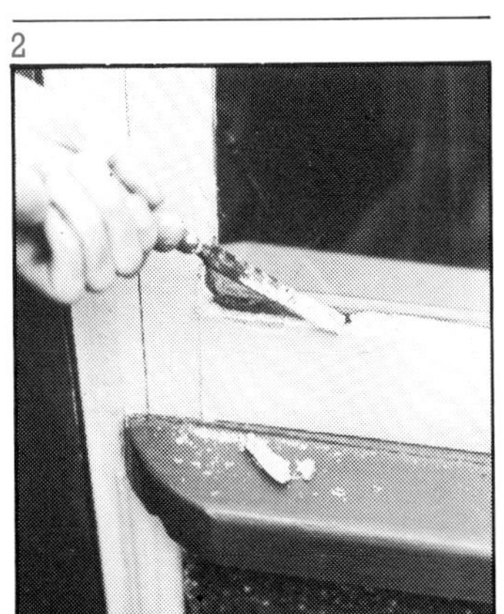

The three basic enemies of timber are wet rot, dry rot and woodworm. Both types of rot establish themselves in wet timber, but wet rot ceases to flourish when the source of water is removed. Dry rot, however, continues to grow and destroy the wood as it can collect water from the air.

The remedy for dry rot is to cut out all infected timber to a point at least 600mm beyond the last visible signs of rot. This timber must then be burned. Infected plaster and other materials must also be removed and the whole area given a treatment of fungicidal solution.

Timber under attack by woodworm deteriorates more slowly, but just the same, it is important that remedial treatment is undertaken as soon as possible, otherwise the amount of work needed to effect complete eradication will be considerable.

The common furniture beetle (generally known as woodworm) is dark brown and about 3 or 4mm long. Its eggs are laid in bare wood such as open joints, cracks and end grain. When these eggs hatch the grubs bore their way into the wood for the next few years. Eventually, they pass through a chrysalis stage near the surface of the timber before boring their way out to leave the well known flight holes.

In certain parts of southern England timber, particularly roofs, can be attacked by the house longhorn beetle. This beetle is some 10 to 20mm long and the flight holes are oval and from 6 to 10mm wide. Attacks by this large beetle are serious and expert advice should be obtained on dealing with them.

1 Floor joists which are half eaten away by woodworm, like this one, need replacement
2 Chipping out rotten putty which is allowing water to get behind the glass into the woodwork. When dry the frame should be primed and reputtied

Enemies at bay!

3 Left untreated, dry rot can become as serious as this example
4 Only simple spray equipment is needed for applying timber preservatives
5 Colourless preservative is ideal for hardwood doors and cottage timbers

Treatment
The treatment for wood attacked by woodworm is to apply a preservative using either a spray or a brush. Spraying to refusal gives better impregnation than brush treatment and it is more effective in reaching otherwise inaccessible areas.

You can obtain fungicidal or insecticidal preservatives and also combination grades. This latter preservative is ideal for treating new timber or previously untreated timber which has not been subject to attack.

These preservatives are generally organic fungicides and insecticides dissolved in white spirit or other medium. There are also preservatives made up into a kind of paste which is spread thickly over the surface of the wood so that it will gradually

3

4

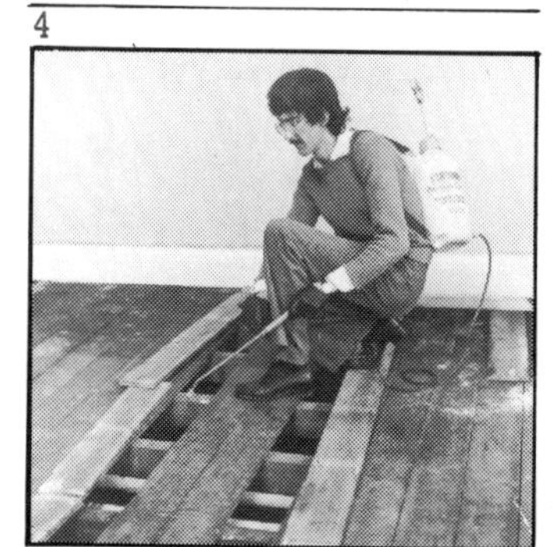

5

penetrate the timber to a greater degree than is possible with spray or brush. Another method of treating timber is to insert pellets of preservative into holes drilled in the wood and these also give off a continual preservative treatment to impregnate the wood.

For spray treatment a pressurised garden spray can be used and it will be very effective in using proprietary preservatives. When undertaking this treatment you should wear gloves and protective goggles to keep any drifting spray out of your eyes. If you are working in the roof or in any other confined area, you should also wear a mask, because it is not good for you to inhale quantities of these liquids.

Always treat all the timber even though it may appear to be sound. The spray will kill off any eggs or fungus spores that are lying on the surface. Any repairs must be carried out using treated timber and cut ends should be dipped in preservative before the wood is fixed into place.

When timber has been weakened by a serious attack of woodworm, the affected timber should be cut out and burned. It can then be replaced with new, treated timber. Slight outbreaks can be treated by squirting preservative solution into the flight holes. These can then be filled with putty or other filler and a close watch kept to see if further holes appear.

At one time, creosote was the main preservative used for timber, and though it offers good protection it does suffer from various disadvantages. It has a strong and persistent smell. It bleeds from the wood and cannot be painted over easily. It is not kind to plants.

There are now, fortunately, proprietary preservatives of various kinds which overcome these traditional problems. You can get a clear type which can be painted over, or you can have a type which will not harm plants. There is a tinted preservative for use on western red cedar which will not only reduce the water absorption to which the timber is prone, it will also restore the faded colour.

Traditional brown preservatives are still available for use on fences and garden sheds, and these will soon weather so that they do not affect plants.

Timber preservatives must not be left on their own to protect the wood. The sources of damp which gave rise to the outbreak must be removed. Faulty or missing damp proof courses must be replaced and leaking pipes and gutters must be repaired.

Keep a sharp watch out for cracks in timber and for open joints in frames as these both will let water into the wood behind the preservative. Any softness, dampness or shrinkage should be investigated and the affected timber cut out, as well as the cause being removed.

The preservation of timber is not a difficult task, it is in general a matter of attention to detail which will prevent a serious deterioration of the wood.

Coming to terms

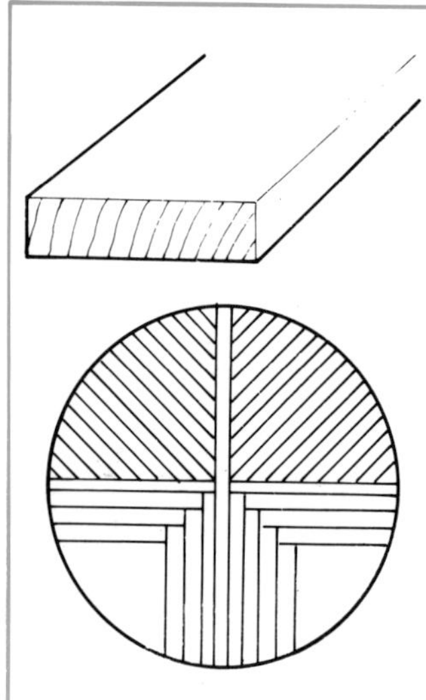
Top: A quarter-sawn board. Above: Two methods of producing quarter-sawn timber

Jargon is with us nearly everywhere these days and the timber trade is no exception. Many of the terms used are understood only by those in the business, but here are a few with which you should become familiar.

Architrave: A moulding which covers the joint between joinery and other surfaces, such as a wall.

Arris: An edge formed by the meeting of two surfaces.

Green: Timber, newly felled, which has not yet been seasoned.

Hardwood: Timber which comes from deciduous trees. Not necessarily hard; i.e. balsa is a hardwood!

Heartwood: Timber in which there are no living cells; it is much harder and more durable than the rest of the tree.

Kiln dried: Timber which has been artificially dried (to a moisture content between 12 and 15%) in a kiln.

Knotting: A liquid, of shellac dissolved in methylated spirit, which seals knots and prevents resin from exuding.

Moulding: Contoured or shaped piece of timber.

PAR: Timber planed all round. It will measure about 4mm less on each dimension than the nominal size.

PBS: Timber planed on both sides only.

Pith: The central core of a tree or branch – made up of soft tissue.

Quarter-sawn: Timber cut from the log so that the circular growth rings meet the surface of the plank at an angle greater than 45°.

RS: Rough sawn timber; measurements will be very close to the nominal ones.

Sapwood: The outer layers of a tree in which there are living cells and reserve foods.

SE: Sawn edge timber (Same effect as PBS).

Seasoning: The drying of timber in air or in a kiln, to make it into a usable material for the woodworker.

Shakes: Splits, caused by the separation of the fibres, along the grain of timber.

Shingles: Thin, short, rectangular pieces of timber, usually tapered, for covering roofs and building sides.

Warp: Bending of timber, usually caused by the evaporation or absorption of moisture.

Imperial or metric?

Although timber has officially been metric for more than a decade, it is perhaps not surprising that many merchants still supply in Imperial measurements. It is simply that customers still work in, and ask for, supplies in Imperial measurements.

The difference

The man behind the counter is aided a little in his seemingly impossible task because the metric measurements are, in general, about 3% less than equivalent Imperial ones.

First, let us look at timber. For our purposes we will consider rough sawn and planed material from 12mm × 21mm (planed) to 50 × 100mm (sawn). What starts out as 16 × 25mm sawn comes out planed as 12 × 21mm – so you lose about 4mm on each dimension during the planing process.

Thickness	Widths
16 (12)	25 (21), 38 (34), 50 (45)
19 (15)	19 (15)
25 (21)	25 (21), 38 (34), 50 (45), 75 (70), 100 (96), 150 (146), 175 (171).
38 (34)	38 (34), 50 (45).
50 (45)	50 (45), 75 (70), 100 (96)

sawn (planed)

Dowel rod, probably the next most useful item, comes in 6, 9, 12, 15 and 18mm diameters.

Boards, of plywood, chipboard and blockboard, come in a wide variety of thicknesses from 4mm to 26mm, and it is here that the metrication programme seems to have gone awry! This is because most boards are supplied in sizes which match, very closely, the old Imperial sizes.

Plywood comes in thicknesses from 4mm upwards and sizes offered go from merchant-cut pieces 12 × 12in. (!) to 2440 × 1220mm. Hardboard starts at $\frac{1}{8}$in or 3mm. thick and dimensions go from 24 × 48in. to the magical 1220 × 2440mm.

Chipboard is probably the most versatile with at least two (Finnish) manufacturers offering sheets 3.2mm ($\frac{1}{8}$in) thick, while most cover the 6 to 26mm range. Again, sheet sizes start around 24 × 36in. up to the normal sheet size of 2440 × 1220mm, while flooring grades are offered in 18 and 22mm thicknesses, 2440 × 610mm. as well as full sheet size. Flooring chipboard can be had either square-edge or tongue-and-groove on two or four edges.

GET CONVERTED

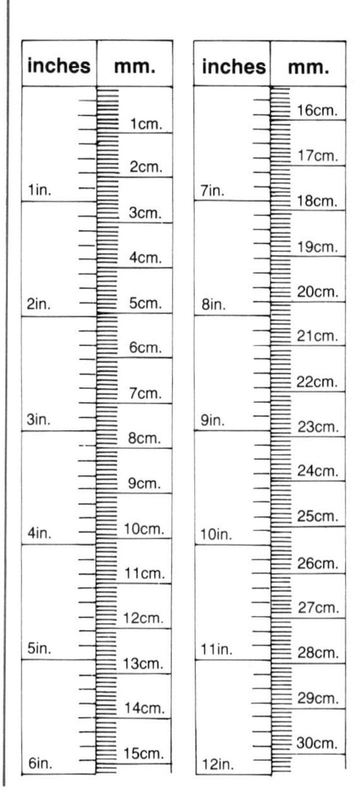

Know your terms

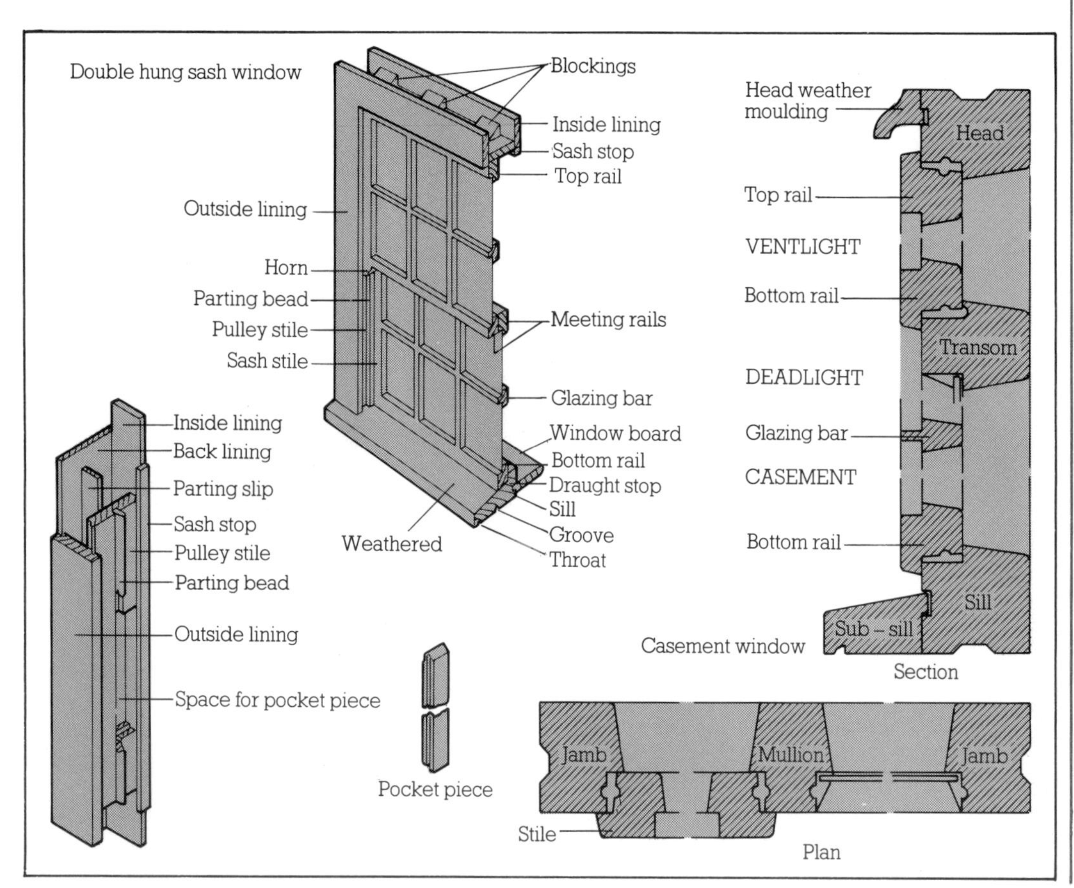

TILES AND TILING

Contents

Square deal

Materials supplied in tile form have always been popular, for so often it is easier to handle and shape small pieces, rather than struggle with sheets of expensive stuff. And it offers a way of producing a variety of patterns with just a few basic units.

There is nothing new to this concept, for ceramic tiles have been traced back to around 4,000 BC in the Near East, while closer to home the Romans knew a thing or two about tiles, both for intricate mosaics on walls and floors, and for tiling their roofs. It seems the Greek and Roman civilisations were using roof tiles at least 2,000 years ago.

But it wasn't until the 1830s and 40s that there was a real growth in the use of ceramic and concrete tiles in this country, until today their production is measured in tens of millions of square metres. And the range of materials in tile form has grown. Anything from mirror and carpet to cork and expanded polystyrene, or parquet and vinyl is available in easily handled packs.

A word of caution

While we have given a fair amount of information of roofwork, we would stress the importance of safety when working at height. See page 72. If you are unhappy with roof work, call in an approved contractor.

Over your head

How is your roof standing up to the winter weather? Is it really fit to withstand ice, snow, wind, rain and perhaps gales? It has been estimated that something like six million pre-1939 homes in this country are potentially in need of re-roofing. But in many cases a patch-up job has been done to extend the life of worn-out roofing.

Making sure your roof is sound is probably number one priority as far as home improvement is concerned, for if the roof fails, most of what has been improved or renovated below is at risk.

There are very clear danger signs when a roof is at risk, and you will find some of these illustrated on page 70. For replacement work, concrete tiles are the most economical form of pitched roofing.

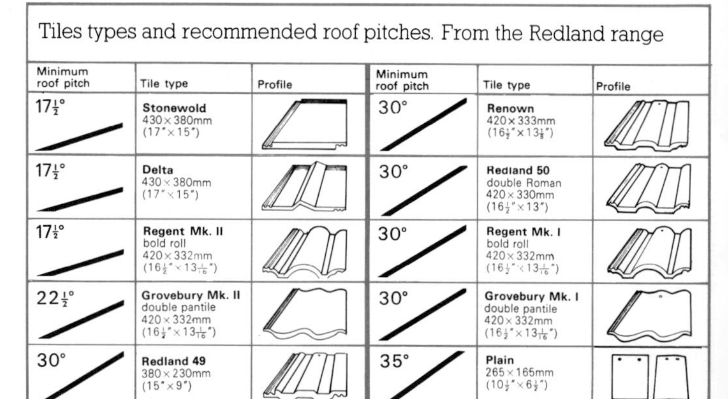

Tiles types and recommended roof pitches. From the Redland range

Minimum roof pitch	Tile type	Profile	Minimum roof pitch	Tile type	Profile
$17\frac{1}{2}^\circ$	**Stonewold** 430×380mm (17"×15")		30°	**Renown** 420×333mm ($16\frac{1}{2}$"×$13\frac{1}{8}$")	
$17\frac{1}{2}^\circ$	**Delta** 430×380mm (17"×15")		30°	**Redland 50** double Roman 420×330mm ($16\frac{1}{2}$"×13")	
$17\frac{1}{2}^\circ$	**Regent Mk. II** bold roll 420×332mm ($16\frac{1}{2}$"×$13\frac{1}{16}$")		30°	**Regent Mk. I** bold roll 420×332mm ($16\frac{1}{2}$"×$13\frac{1}{16}$")	
$22\frac{1}{2}^\circ$	**Grovebury Mk. II** double pantile 420×332mm ($16\frac{1}{2}$"×$13\frac{1}{16}$")		30°	**Grovebury Mk. I** double pantile 420×332mm ($16\frac{1}{2}$"×$13\frac{1}{16}$")	
30°	**Redland 49** 380×230mm (15"×9")		35°	**Plain** 265×165mm ($10\frac{1}{2}$"×$6\frac{1}{2}$")	

To anchor a loose slate, nail a strip of lead or zinc to a convenient batten

Slide in the slate and bend up the strip to hold it firmly in place

As extra security put a blob of epoxy-based adhesive under the strip

1

2

3

4

1 Neat re-roofing job
2 A new roof with plain tiles for the porch
3 Ventilators ensure a good natural flow of air
4 Vertical tiling has an important role in renovation work

Over your head

Here are some danger signs to look for:

1 Defective flashing allowing rain in

2 General breakdown of tiles due to severe weathering

3 Deterioration along valleys

4 Missing and displaced slates, and too many visible repairs

5 Repair strips not strong enough to hold slate in place

6 Dislodged capping tiles and damaged flashing

1

2

3

4

5

6

Roof Types

The three most likely forms of roof construction you will encounter are illustrated here, and as you will see, all three are based upon battens to which the roof covering is nailed.

The underlay, again common to all three, plays an important role in keeping out in-blown rain and snow and preventing too much fresh air entering the roof space. Ventilation is important, and companies like Redland and Marley have systems designed to provide adequate ventilation while at the same time maintaining the weatherproofing qualities of the roof. The underlay goes on to the rafters with not less than 100mm lap horizontally and vertically, except on pitches below 35° where not less than 150mm horizontal lap is required. Fixing is with clout nails.

A roof clad in plain tiles is illustrated in fig 1. The traditional plain tiles have a double overlap and therefore involve more weight, greater labour in laying and lead to a reduction in effective pitch. This makes the whole roof a more expensive proposition.

With single lap tiles, as illustrated in fig 2, you will see that the overlap is considerably less. So less weight – and expense – is involved, and fixing is less labour intensive.

Slates, fig 3 are durable, but a slate roof usually fails because the fixing nails have rusted through. Slipping slates can be re-fixed as shown on page 4, but where large areas of loose ones are involved, complete replacement will be advisable. Reslating would prove very expensive. If you catch a failing roof early, there are systems of anchoring slates from within the loft space using a special adhesive compound.

Wherever nails are replaced in roof work, or where new work is undertaken, be sure to use rustless ones. These may be galvanized or alloy.

Small wedges will help you remove damaged tiles

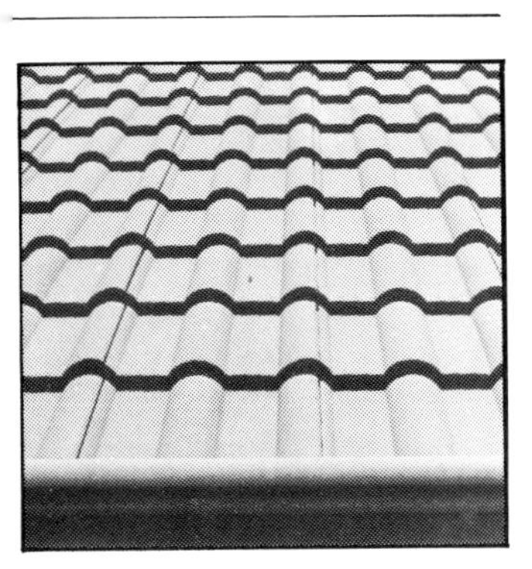

1 2 3

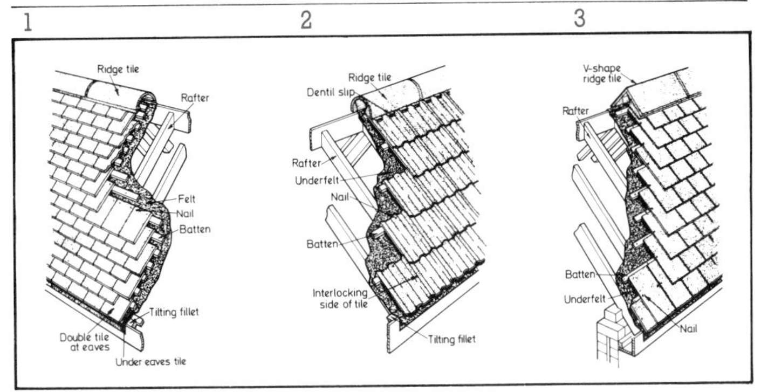

Over your head

Safe access

Working at roof level calls for the right equipment. You will need an extension ladder which gives you at least three clear rungs above gutter level so there is always something to get hold of when getting on and off. It should be anchored so it can't slide or slip top or bottom. A sandbag will do at the base, and a ring bolt into the fascia board to which the ladder can be roped at the top.

For work on the roof you need a roof ladder of the type which can be wheeled up the slope, then turned over to hook over the ridge. This is an item you could hire if you are not likely to need it often. Or you can buy an attachment.

Steel roof

A totally different approach to roofing is offered by Plannja International with a system called Scanroof.

The roofing comes in strips and is made of steel 100% corrosion-proofed. The strips are formed to resemble tiling, but with this method no battens or plywood deck is required. The strips can be anchored direct to the rafters, making the build up of a roof simple and fast. Because the units are light in weight it also simplifies handling, and once complete needs no further maintenance. Details of the system can be obtained from the address given on our reference page at the end of this magazine.

Home and dry

1

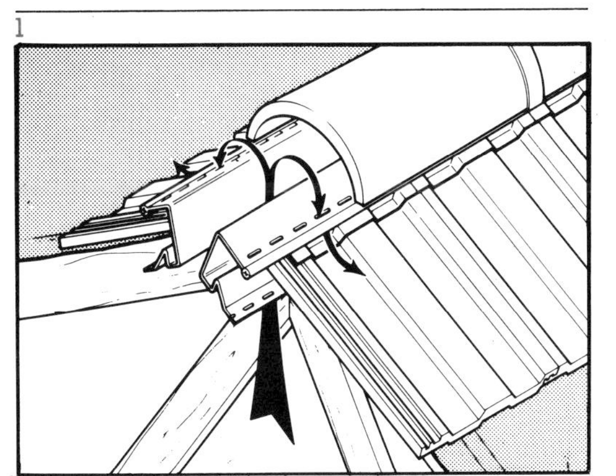

2

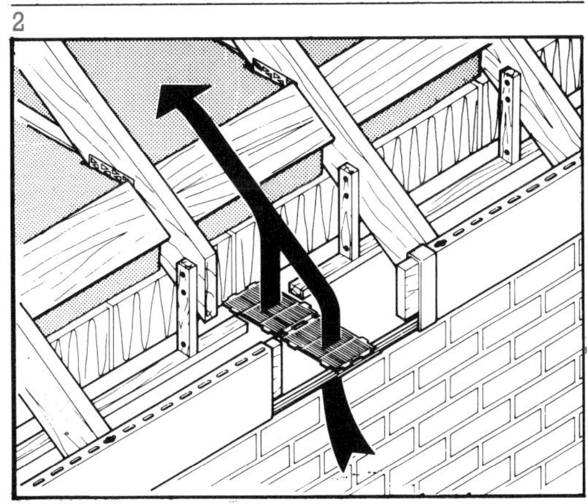

3

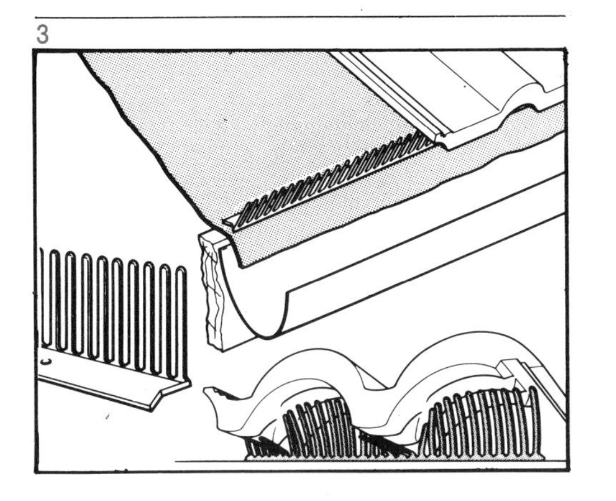

4

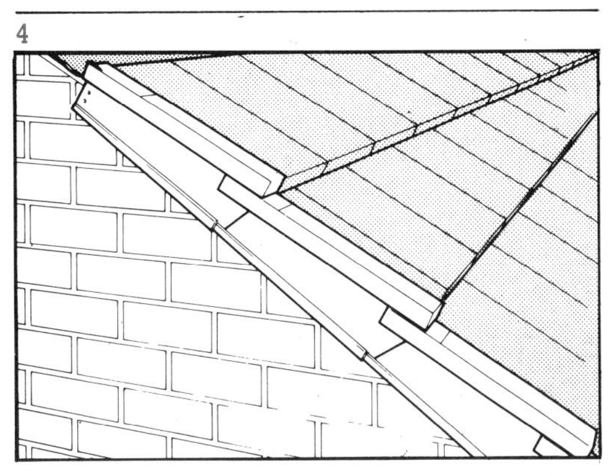

A complete dry-fix roofing system, fully meeting the Building Regulations and proved in severe weather, has been introduced by Marley Roof Tile Co Ltd. There are four main components. A ventilated dry ridge, fig 1., designed so that moist air escapes through slots in the ridge batten. A Upvc fascia/bargeboard and ventilated soffit, fig. 2., which allows the passage of air through a ventilation grille in the soffit. A neat comb eave filler fig 3, designed to fill those awkward voids which the birds love. The comb pattern allows for ventilation while at the same time taking up the contours of the tile profile. And finally a dry verge system, fig 4., which uses interlocking units adjustable for tile laps between 75mm and 115mm.

We would stress that it is not a d-i-y system.

Tiles around the house

Tiles can play a part in the décor of every part of the house. Here are a few examples

1 Thanks to modern materials, carpet can be supplied especially for kitchen use.

2 Pretty tiles for this dressing table

3 Tile effect with cork background

4 Attractive textured ceiling tiles

1

2

3

4

5

6

7

8

9

5 A bathroom 'doubled' in size with stylish mirror tiles
6 The blue theme carried on to work tops and sink surround ties units together
7 Parquet tiles make an ideal surround for a carpet
8 Tiles and timber blend well together in this modern bathroom
9 Mirror tiles used to good effect in a small bedroom.

A feeling for ceilings

TILING A CEILING

Expanded polystyrene tiles are ideal for improving a jaded ceiling, and for hiding minor cracks and blemishes. You will find they come in a number of sizes, and the larger the tile the neater the effect.

If you plan to emulsion paint the tiles, remember the ideal time to do them is before you put them up. It is far less tiring, and you can be sure to paint all the edges well. Remember you must not use gloss paint, for while all expanded polystyrene tiles are now self-extinguishing, the addition of gloss paint will allow them to hold a flame should there be a fire.

Apply adhesive to the whole ceiling, having made sure it is clean and grease-free. The old five blob method we used to use is not allowed, for should there be a fire the blobs allow molten pieces to fall away from the ceiling. An overall coating of adhesive does not. Cut tiles to size with a sharp craft knife and straight-edge. There are no general rules for placing ceiling tiles, and often it pays to start where the tiles are seen most, such as at a doorway, then do the cutting where the ceiling is seen least.

Obviously, cut edges may need a touch of paint.

Don't apply too much finger pressure to textured tiles or marks will remain.

1 An attractive patterned ceiling in flame-retardent expanded polystyrene tiles

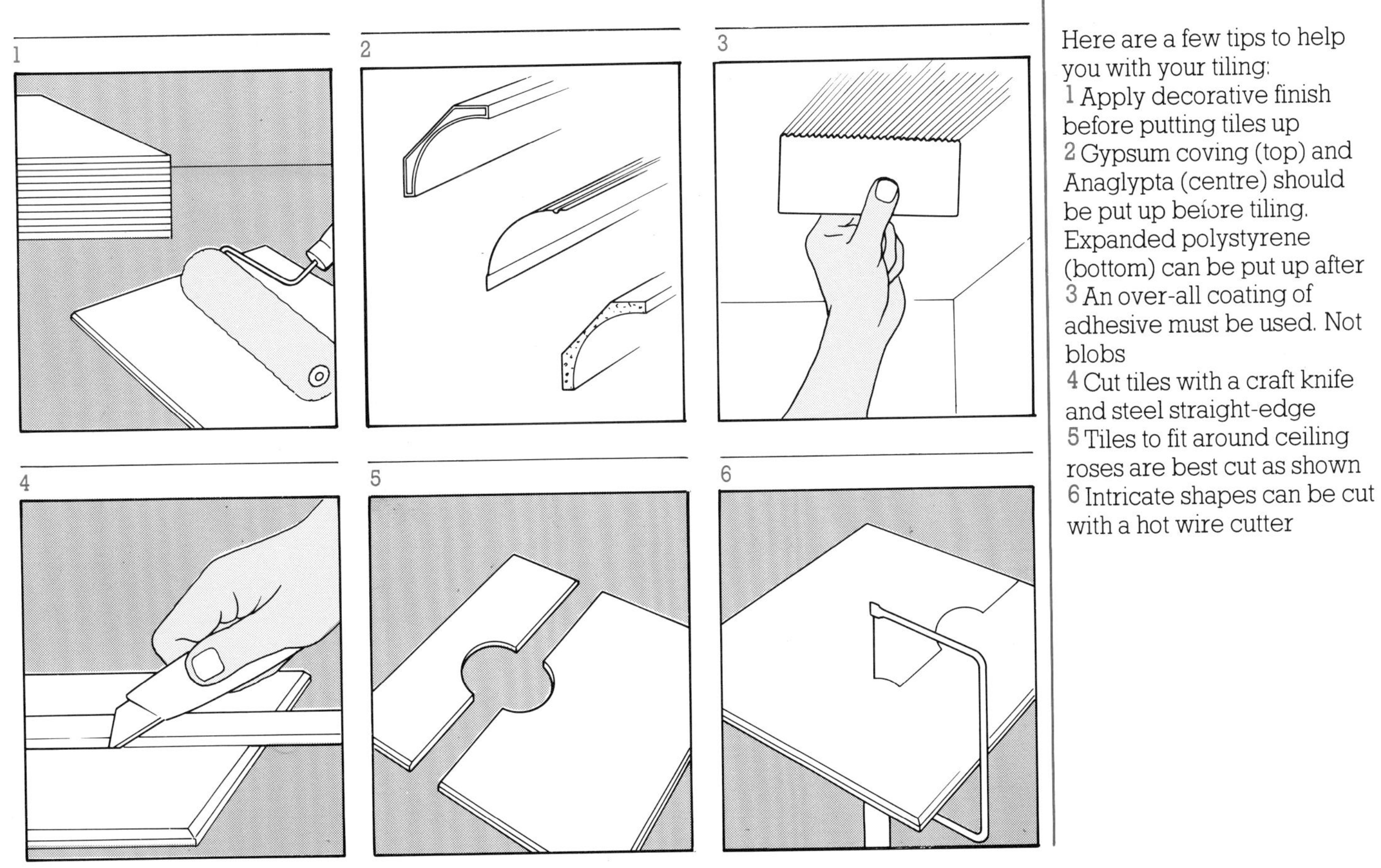

Here are a few tips to help you with your tiling:

1 Apply decorative finish before putting tiles up

2 Gypsum coving (top) and Anaglypta (centre) should be put up before tiling. Expanded polystyrene (bottom) can be put up after

3 An over-all coating of adhesive must be used. Not blobs

4 Cut tiles with a craft knife and steel straight-edge

5 Tiles to fit around ceiling roses are best cut as shown

6 Intricate shapes can be cut with a hot wire cutter

A feeling for ceilings

1 Gaps between boards need filling with joint filler ...
2 ... then paper tape is added to disguise the joint

CEILING REPAIR

While ceiling tiles are ideal for covering minor blemishes, if a ceiling is in poor shape it will pay to do some remedial work before tiling. And remember this is the ideal time to put in any new wiring for lights, as you can hide the work under the tiles when you've finished.

In older houses you may find the old lath and plaster has been used for ceiling construction. And during the war years many of these old ceilings were shaken up and some lost their keying. If you find a bulge, it may be possible to push it back in place and secure with rustless screws. Or, if you can get at the ceiling from above by lifting floorboards, you can form a new key by pouring plaster of Paris on to the laths. As a last resort the whole lot can be taken down, but this is a messy job.

Modern plasterboard gives little trouble, though you may have cracks at joints if too much weight is applied to a loft ceiling. Joints need strengthening with cotton scrim and filler to try to restrict the movement of the joint. And you need to ease the weight above.

If you encounter stains on ceilings – perhaps from a previous plumbing leak, from in-blown rain or snow or perhaps from an over liberal coating of wood preservative – it is wise to seal them off. Use an aluminium primer sealer on the ceiling. This has a scale-like nature which will isolate the stain so you can decorate over it with no fear of bleeding through.

Should you encounter old distemper on a ceiling, rub it off. Ceiling tile adhesive will not grip to distemper, and your tiles will float down.

1

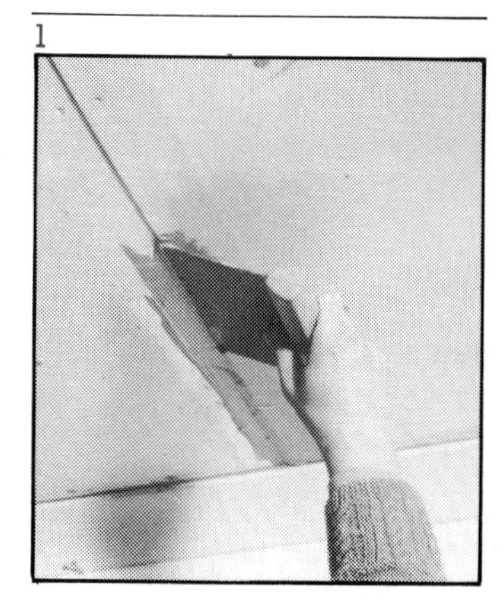

2

Ways with walls

Tiling is a simple and very satisfying job, but like most decorative work it pays to spend time planning – especially if patterns are involved.

If you want an overall plain background effect with some patterned tiles, a ratio of at least five plain background tiles to one patterned tile is recommended for best results. To help plan your design lay out the tiles in a suitable large area and move them around until you get a pleasing pattern.

EXACT SEQUENCE

If you then transfer your chosen design on to paper you'll know the exact sequence when it comes to fixing. You will then be able to plan the design around windows, doors and fixtures. (Do not mark the back of tiles with a marker pen, as this may seep through and stain the tiles when they are fixed. Marker pens may be used on the glazed surface).

When gathering up the plain tiles, mix them to ensure a pleasing overall effect will be achieved. All tiles vary slightly in colour so it's important to mix the contents of the packs.

Now is also the time to select tiles to be used for external corners and edges. Within any pack of Cristal Universal tiles a proportion of tiles will be glazed on one edge or two adjacent edges. Separate these tiles to use on corners and edges. Remember it is inadvisable to place patterned tiles in corners or on edges where they may have to be cut.

Wall tiles can be used on any hard flat surface. Ideal surfaces are existing tiles; well plastered walls; plasterboard; wood – providing it's a rigid surface – and plastic laminate.

Any previous wall covering (not tiles) such as wallpaper must be removed. When tiling on to paint, be certain the surface is well bonded.

Insulated wall tiles

PREPARATION

Just like wallpapering or painting, a little careful preparation is never wasted. Do make sure that the walls to be tiled are free from dirt and grease. If the surface of the wall is damaged in places, fill with proprietary filler, and smooth when set.

No matter if tiling a whole room, one wall, or simply a couple of rows above your bath, it is important that the bottom row of tiles is level. Floors are often uneven. So first establish the lowest point on the floor or skirting board. If tiling only one wall, determine its lowest point. When tiling more than one, find the lowest point of all walls.

To do this, lightly tack a lath to a wall approximately one tile high from the floor or skirting. Use a spirit level to get the lath perfectly horizontal. With a ruler find the point at which the floor is at its greatest distance from the lath. This is the lowest point of the floor. (If you are tiling more than one wall, continue the lath around the adjacent walls at exactly the same height, and check that there are no lower points).

Having found the lowest

Ways with walls

1 Lay out the tiles on the floor so you can play with the pattern
2 These are the tools you will need
3 The cutter should just whisper over the tile
4 Break the tile over a matchstick
5 Score any area to be cut away . . .

Whether you are tiling the kitchen or bathroom, the basic principles are the same. We are grateful to H& R Johnson (Cristal) for this sequence of illustrations showing how the job should be done

1

3

2

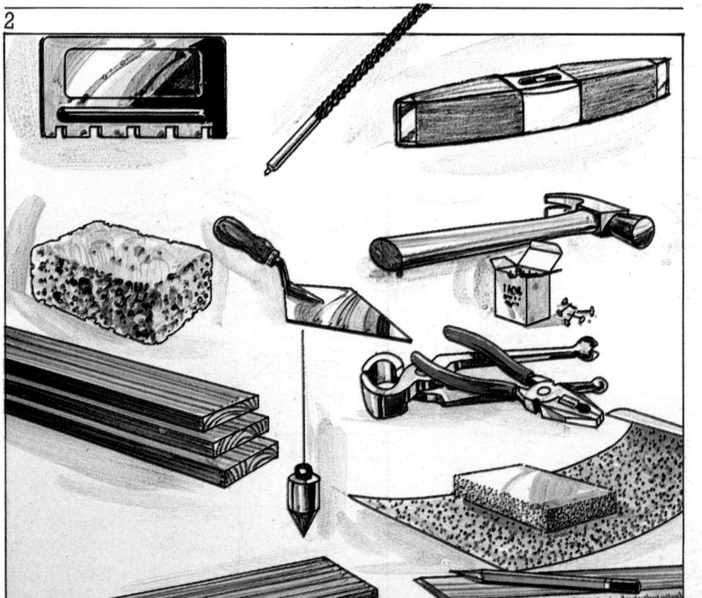

4

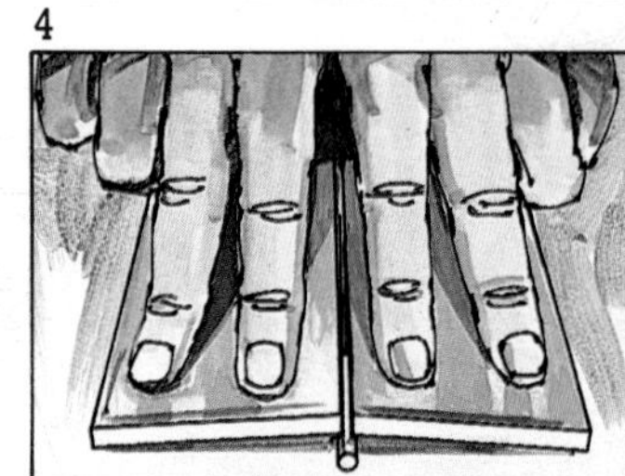

5

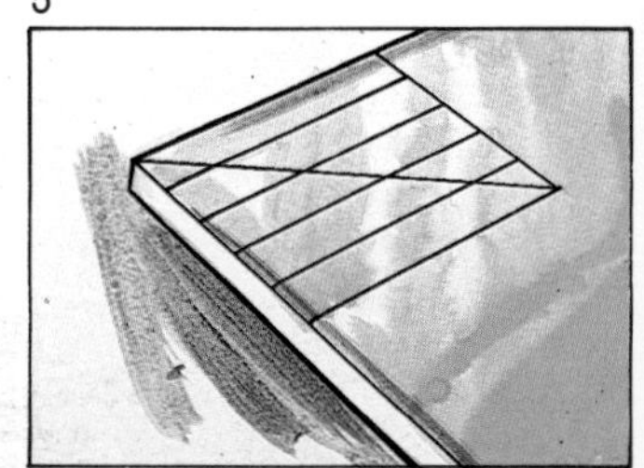

6

7

8

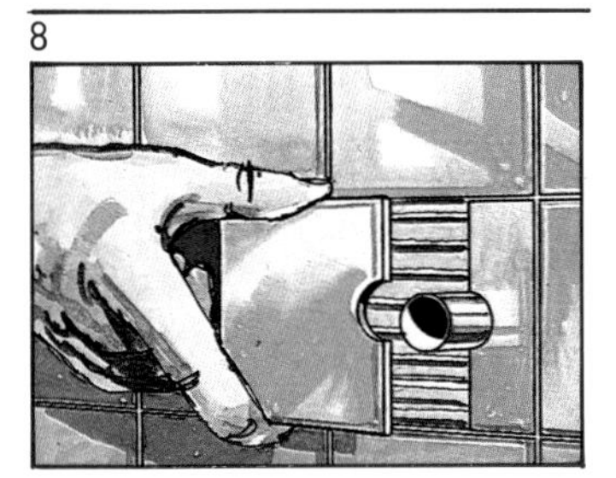

9

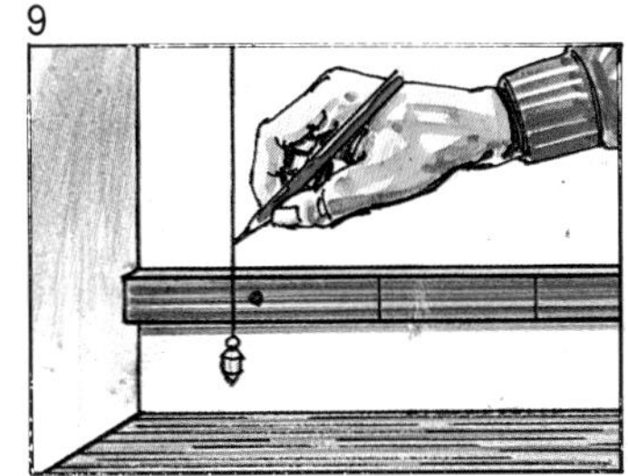

10

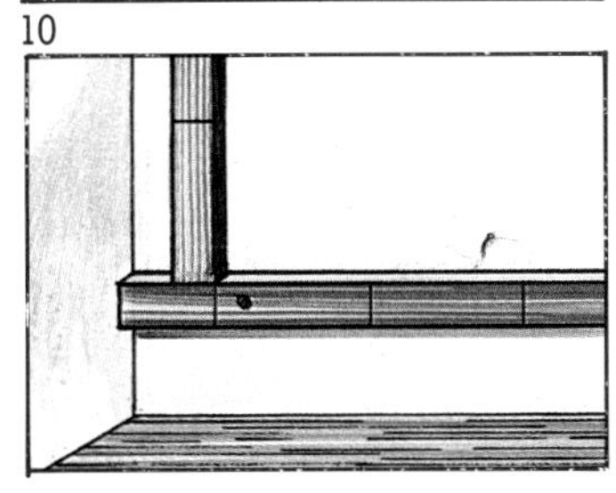

11

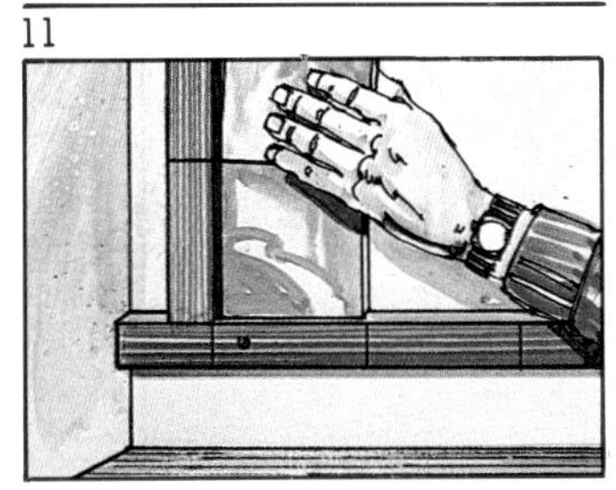

12

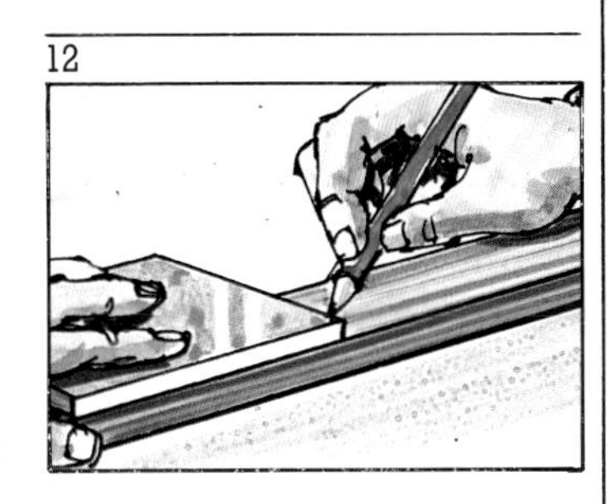

13

14

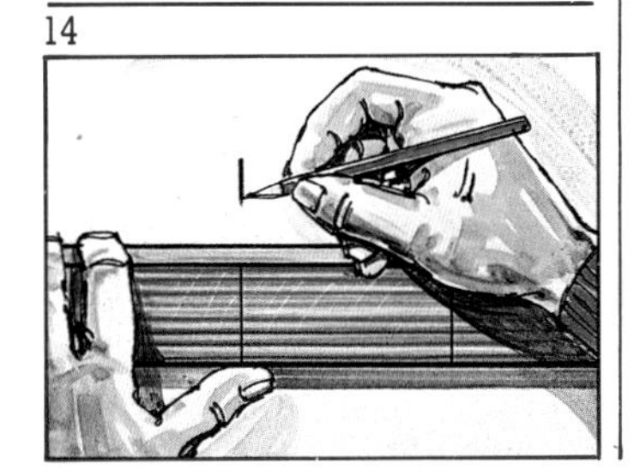

6 . . . then nibble the waste away with pliers (protect your eyes!)
7 The same applies to cutting a curve
8 This is how to cut a hole in the middle of a tile
9 Check for vertical as well as horizontal . . .
10 . . . and fix a vertical lath
11 This is where you start tiling
12 Where there are features to negotiate, make up a measuring staff
13 Use this to centralise the position of your tiles around any major focal point so you get equal cuts each side
14 Mark the wall to indicate where the outside edges of the cut tiles will appear

point, refix the lath firmly to this wall with its upper edge one tile high above the floor or skirting at this point. If tiling more than one wall, continue your laths at this height across the other walls, using a spirit level to ensure all laths are perfectly horizontal.

Usually all tiling jobs involve cutting tiles, either around windows or wash basins, or at either end of the wall. Before planning a layout, decide what is the major focal point within the room. It is important that tiling is arranged symmetrically around the focal point.

For example, on a wall without features you will require an equal cut tile at either end of the wall. If the wall contains a feature such as a window you will want an equal cut tile either side which may necessitate unequal cut tiles at the ends of the wall.

Equal cut tiles can be achieved by using two methods. Method A is recommended for walls free of features like windows or wash basins. Method B for walls with features.

METHOD A – FOR WALLS WITHOUT FEATURES

1. Place a row of tiles on the floor close to the wall to be tiled, ensuring they are placed edge to edge.
2. When you've placed as many tiles as you can along the length of the wall move the row of tiles to the left or right until you have an even space at either end as near to half a tile as possible.
3. To establish a vertical line to one corner of each wall, mark on your lath the positioning of the first full tile in your row of tiles. Pin your plumb line near to the top of the wall so that it passes through the mark on your lath nearest the corner.

Draw a pencil line on the wall behind the plumb line, and nail a lath along this line to act as your vertical guide. Check the angle at the intersection of the horizontal and vertical laths by placing some tiles loosely in position. They must sit perfectly square.

PLANNING LAYOUT METHOD B – FOR WALLS WITH FEATURES

1. Take the 4 ft piece of straight wood and, starting from one end, mark a tile width and then continue along the wood. This will provide a measuring staff.
2. Use the measuring staff to help you centralise the position of your tiles around the major focal point in order to achieve an equal cut tile at either end.
3. Mark the wall with a pencil to indicate where the outside edges of the cut tiles will appear.
4. To establish your vertical line take your plumb line and pin it to one of the marks you have just drawn showing the edge of the cut tiles.

Make a mark on your lath where the plumb line crosses it. Mark tile widths from this point along the lath until the corner is reached.

Re-position the plumb line in the corner of the room so that it passes through the point on the lath which indicates the edge of the last full tile. Pencil a line on the wall where the plumb line falls and nail up a lath along the line.

Check the angle at the intersection of your horizontal and vertical laths by placing some tiles loosely in position. They must sit perfectly square.

Ways with walls

1

2

3

4

5

6

When tiling with cork:

1 Work from a true horizontal and vertical start point

2 Cut tiles with a sharp craft knife, using a steel straight-edge

3 To cut fill-in pieces, position a new tile over the last fitted tile, then use spare tile as shown. Mark for cutting on new tile

4 Place tiles like this. Don't slide them, or adhesive will come up between joints.

5 To fit around obstacles, make a simple template . . .

6 Then transfer the pattern to your tile

Beauty underfoot

1 For vinyl tiles, your sub-floor must be perfectly smooth, otherwise irregularities will show through
2 Individual tiles can be lifted and cleaned

1

2

Floorcoverings in tile form have much to commend them, and what started off as a very basic and utilitarian idea in the early days of d-i-y has now developed into a really sophisticated market.

The main advantage of the tile is that it is easy to calculate just how much you require with little chance of waste. Handling is greatly simplified, as it takes considerable experience to manhandle a 12ft wide piece.

Another spin-off is that should an area be damaged, only one or two tiles may need lifting and replacing. This is not so easy with vinyl and cork, but with carpet tiles, which are loose-laid it is simple to lift one and clean it or replace it. And with carpet tiles you have the option of lifting tiles which are getting more than their fair share of wear, and changing them with ones in less used areas.

Tile types

Parquet tiles offer a neat and simple way of transforming a floor, and tiles usually consist of a number of separate strips of wood held together on a bitumen backing. Where necessary, strips may be broken off to allow tiles to fit. Tiles vary a lot in quality, but more attention should be paid to the actual neatness of construction and the squareness of the product rather than the total thickness. With most of the hardwoods, a tile as thin as 3mm, laid well and sealed, would last more than a lifetime. A tile out of true will cause laying problems.

Tiles may come pre-sealed, or they may be untreated. In the latter case it is wise to get a coat of seal on as soon as possible after laying so the timber doesn't

1

2

3

1 The attractive warmth of cork
2 Justin Blue, from the Dolly Mixtures collection
3 Parquet may come pre-sealed, ready for traffic
4 Cut pile carpet tiles
5 No problems clening tiles
6 Gemstone ceramic tiling

4

5

6

Beauty underfoot

3 Carpet tiles are easily cut from the reverse side with a craft knife. Fit should be tight
4 Nothing to stop you making up attractive patterns.
5 Realistic brick effects can be obtained in cushioned vinyl.

get dirty or covered in scuff marks.

Parquet may be used as an over-all floor, or be used as an attractive surround to a carpeted area, forming a shallow well in which the carpet can be laid.

Vinyl tiles have improved tremendously over the years, though it should be said you still get what you pay for. In the early days of vinyls, the tiles were laced with quite high quantities of filler, including asbestos, which made them brittle. The modern vinyl tile has a much higher vinyl content and is therefore more flexible and long-lasting.

The only possible disadvantage of a nice flexible tile is that it will follow any irregularities in the floor surface, so it is vital to make sure your floor is smooth and free from projections. See page 88. The warmer your rooms, the more flexible the tiles, so it pays to store them in the room in which they will be laid prior to fixing. If there are undulations in a floor, cover the whole floor with standard hardboard so that a new smooth surface is presented. It will add considerably to the life of the tiles.

Old vinyl tiles often prove difficult to lift. The simplest way is to soften them with controlled heat. Either a domestic iron laid on a tile with kitchen foil in between iron and tile. Or try the new hot air paint stripping gun, making sure you don't overheat the plastic.

Scuff marks on vinyl tiles can be removed with fine wire wool and turps substitute. Only rub lightly.

Cork tiles have an attraction all of their own, and with natural materials being much in favour at the moment their popularity seems to be on the increase. Cork is a natural insulator, so it is warm to the feet, making it ideal for

3

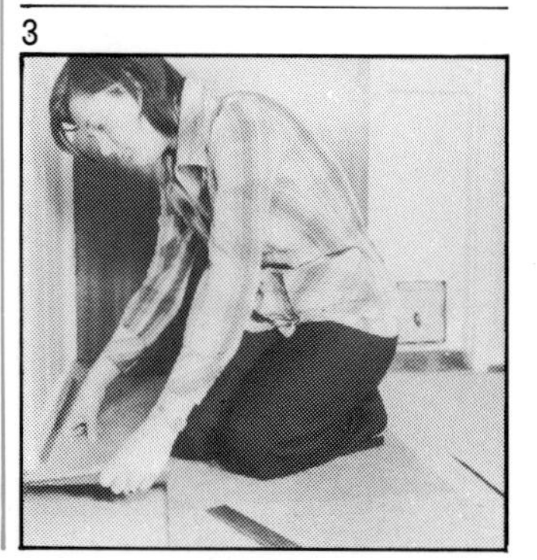

4

kitchen and bathroom use. But in such areas it is important that the cork is well sealed. This may have been done during manufacture, or it may be up to you to treat them when you've laid them. Polyurethane seal is ideal. A priming coat first to seal the pores, then decorative coats to follow. You have a choice of matt or gloss finish. The matt tends to show off the grain of the cork to better advantage. A high gloss finish can be reduced by light rubbing with fine wire wool, but take care not to reach the cork.

Tiles are easily cut as long as you use a really sharp craft knife. A blunt knife tends to rip pieces off – especially at the corners. When laying your tiles in adhesive, place them rather than sliding. The sliding action pushes adhesive up on the face of the tiles. Be fairly liberal with adhesive as cork is rather porous on the under face.

Carpet tiles are growing rapidly in popularity, and they are now available in a number of qualities, suitable for light, medium or heavy wear, and often in an option of sizes. Tiles have a special lay-flat backing so that they need no anchoring to the floor. They are laid so that they press against each other just enough to lock them in place, yet still allowing individual tiles to be lifted out.

Unlike other floor tiles where it is usual to start at the centre of the room and work out, tiling starts against the most prominent wall – such as that incorporating the door to that room – and you work away, pushing each tile against its neighbour. You will see that most tiles have arrows on the reverse side, and it is recommended that these be laid in box formation so that piles in adjoining tiles are always at right angles to each other. This gives the popular chequerboard effect, but if you want a broadloom effect, there's nothing to stop you having all the arrows facing the same way.

If you wish, a little double sided tape can be used to hold cut pieces in place. Or in doorways if there is a risk of the tiles being lifted. The kitchen was never looked upon as the ideal place for carpet, but you can now buy carpet tiles especially developed for kitchen use. Should they become soiled, they can be lifted and cleaned, and the fibres will not hold debris, or stain.

Ceramic tiles grew in popularity with the advent of the packaged holiday. More and more home owners realised how attractive floor tiles could look as they saw them widely used in Spain, Italy and the south of France. At one time they would have been cold to the touch in British bathrooms, but with the increased use of central heating the days of cold floors have gone.

Normally ceramic tiles would be laid on concrete floors where a special floor-

5

Beauty underfoot

ing adhesive would anchor them. But it is possible to lay them on timber floors as long as conditions are right. We consulted AG Tiles who say:

'Your surface must be rigid, flat, dry and free from dirt and grease. Make sure that timber floors are strong enough to support tiles without flexing, and are well ventilated below. You must strengthen floorboards with an overlay of ½in. (12mm) exterior quality plywood screwed (not nailed) at 12in. (300mm) intervals.

'When tiling over timber surfaces, brush a primer such as BAL-bond liberally over the whole of the area to be tiled. Allow it to become touch-dry before spreading the adhesive, and remember to clean your brushes in water immediately after use.'

Concrete surfaces to be covered also need to be flat and free from projections. See below for details.

Be warned that ceramic floor tiles are not as easy to cut as wall tiles. If you plan to tackle a whole floor you would be wise to borrow a professional tile cutter which scores and strikes the tile correctly, ensuring a true cut. Even so, it is wise to allow 10% extra for cutting and shaping.

Flooring problems

Tiles, of whatever kind, should never be used to cover a floor in poor condition. Very quickly faults will show through, and projections, like concrete nibs or old floor tacks will soon damage your new floorcovering. So, spend time ensuring that the surface you are covering is clean, dry and flat.

With timber floors, make sure that all the boards are firmly secured, and that there is no uneven surface – such as where old boards are worn. If a floor is uneven, hire a floor sander and, after carefully removing all loose tacks and punching down proud nails, re-surface the floor. You may need a small belt sander for areas where the larger sander will not reach. It will have a suction device to transfer most of the dust to a bag.

If you encounter woodworm, check to see there is no infestation below the boards, then treat the boards with woodworm killer.

If you have concrete floors remove all projecting nibs which could damage your new covering, and fill any hollows with screeding compound. Where a floor is uneven all over, it can be re-surfaced with the screeding compound. With the modern materials, no actual trowelling is needed, so application is simple.

If there are signs of damp in the floor, it would be wise to treat the whole floor with a bituminous waterproofing compound. If there is serious rising damp, get expert diagnosis. Where concrete is dusting up, coat it with pva adhesive diluted 1 part adhesive to 2 parts water.

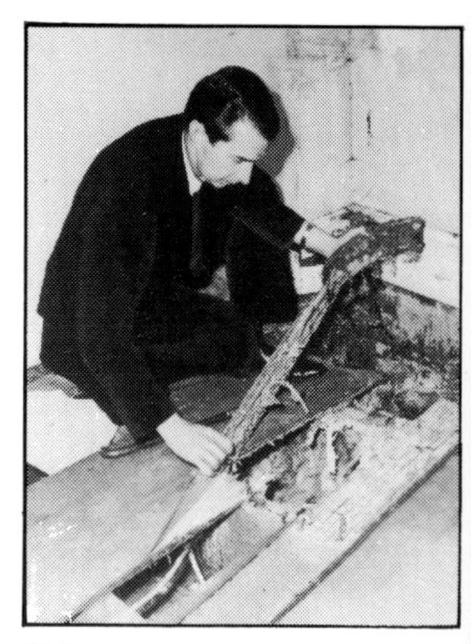

If dry rot gets hold of a floor, results can be catastrophic, as this Rentokil photo shows

Is this your problem?

Tiling around the home can be an enjoyable and satisfying task, whether it involves walls or floors in kitchens or bathrooms, worktop or table top surfaces, or exterior floors, patios or porches. But there are some important points to watch in order to ensure success. The following topics ure those most commonly raised by users of BAL products. Perhaps they are your problems too.

The finished tile installation will look much better if you avoid unsightly cut tiles and ensure that joints are even in width. Therefore, it pays to set out the tiles before you start fixing. With floor tiles, for instance, mark the centre lines of the floor with chalk fig. 1 and place 'dry' tiles along each line from wall to wall, fig 2. If the gap between an end tile and the wall is too narrow, remove one tile and adjust the row until the gaps at each end are of equal width. Ideally, the gaps should be a quarter of a tile width or more, as anything narrower than this could prove difficult to cut. Floor tiles with spacer lugs will automatically ensure equal joint widths. For non-spacer tiles, allow 3mm (⅛in) between joints.

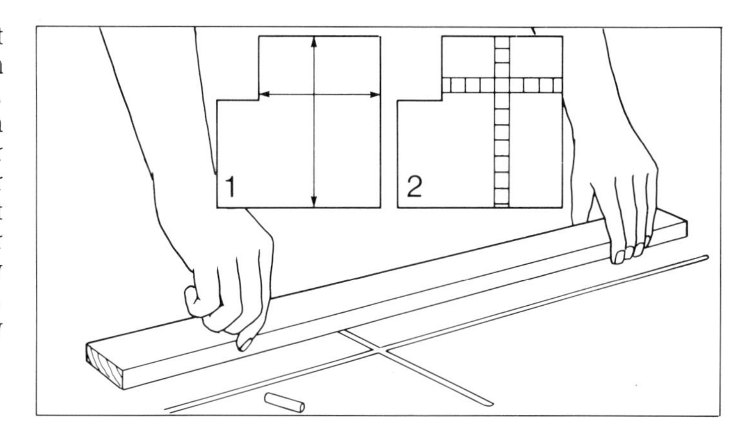

Selecting the correct adhesive

There are many different tile adhesives on the market but you need to use one which is correct for your particular job. For instance, if you are intending to lay tiles on an external patio floor you will need an adhesive with good water resistance. For interior situations this will not be required unless the tiles are intended for a shower compartment. Again, different types of wall surface call for different types of adhesive. So how do you decide what to do?

Fixing wall tiles to plaster surfaces

Key points are as follows:–

1. Plaster must be at least four weeks old, clean, dry and flat.
2. Tile only on to a FINISH coat.
3. Check condition of plaster (not friable or dusty).
4. Priming may be necessary especially on lightweight plaster surfaces (e.g. with BAL-Primer) and the weight of tiling should be kept to a minimum (e.g. use tiles not exceeding 6.5 mm in thickness).
5. Plaster is not recommended in wet areas e.g. shower compartments.
6. Cement-based adhesives are not recommended for

Is this your problem?

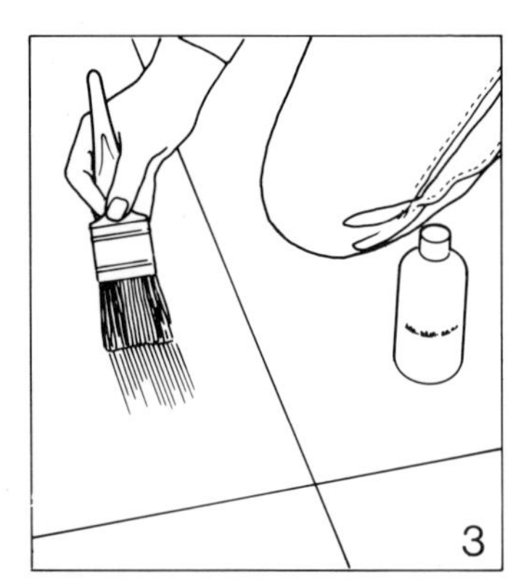

use on plaster backings. Instead ready-mixed, organic-based adhesives should be used such as BAL-Wall or BAL-Grip.

Fixing wall tiles to sheets and boards e.g. plasterboard, plywood

Key points are as follows:–

1. The sheets or boards must be adequately braced.
2. Paint or seal exposed edges and back of sheets or boards but not the surface to be tiled.

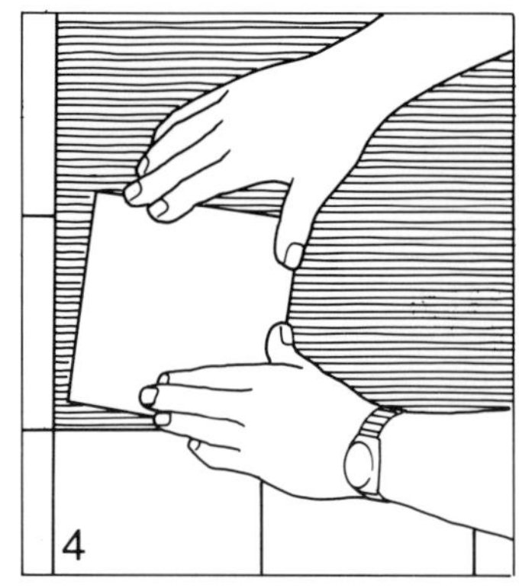

3. Check condition of surface.
4. Use ready-mixed, organic based adhesives e.g. BAL-Wall or BAL-Grip or two-part adhesives such as BAL-Flex for fixing tiles.

Use of BAL-Cem adhesive for laying tiles on other types of base

With BAL-Cem, ceramic floor tiles can also be laid over wood, firmly-fixed vinyl tiles, existing ceramic floor tiles, quarries and terrazzo.

Key points are as follows:–

1. The floor to be tiled must be rigid, flat, dry and free from dirt, grease and polish.
2. Make sure that timber floors are sufficiently strong to support the tiling without excessive deflection and have adequate ventilation underneath.
3. Suspended timber floors can be strengthened by an overlay of ½in (13mm) plywood screwed at 9in (225mm) centres.
4. When tiling over existing vinyl, quarry tiles, etc. ensure that they are securely bonded before laying tiles.
5. Timber surfaces and vinyl tiles must be primed with BAL-Bond Priming Agent before using BAL-Cem. Brush BAL-Bond liberally over the whole of the surface to be tiled and allow to become touch-dry before starting laying. See fig 3.
6. Do not put the floor into service until the adhesive has set firmly and the tiles are free from any tendency to move. Do not walk on the floor for at least 24 to 48 hours.

Wall tiles in dry situations

Use the notched trowel methods. This means that you spread the adhesive on the wall and comb with a notched spreader. Press the tiles home with a slight twisting action – see fig 4.

Exterior walls

1 You can see here the state of the old tiling
2 For safety, a guard rail was used on the scaffold
3 Old tiles were easily ripped away with a claw hammer

1

2

3

While considering a feature on vertical tiling, we were fortunate to be invited to see some work being done on a house which was built in 1939. It was located near Sevenoaks. The vertical tiling was in a poor state, so Marley were involved a local contractor in stripping then re-tiling with Marley plain tiles.

The owners will be forever grateful for their decision to re-tile for when the old ones were removed, it was found that apart from a layer of feather-edge boarding, there was nothing between the interior plasterboard walls and the outside air. They said the rooms were bitterly cold in winter.

With all the old cladding off, all gaps between timbers were stuffed with mineral wool blanket providing a superb insulating layer over which bitumen underlay was

Exterior walls

4 Removal of the old boards revealed lack of insulation
5 The internal plasterboard was clearly visible! No wonder the bedrooms were cold in winter
6 All cavities were stuffed with mineral wool blanket
7 Underlay was then secured, with a generous overlap
8 Chalked string helps mark the position of battens
9 Battening taking shape between windows

4

5

6

7

8

9

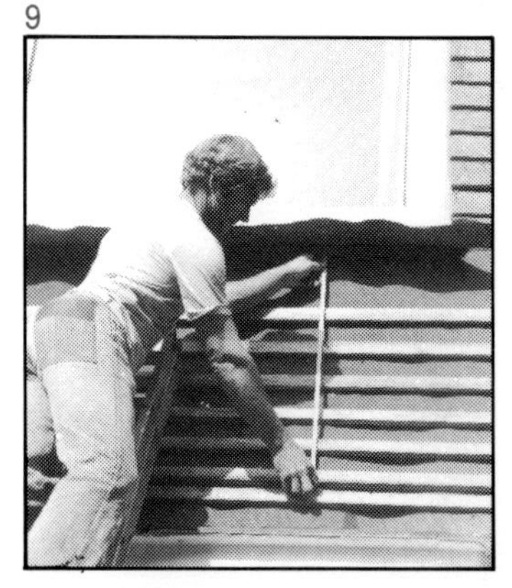

10

11

12

13

14

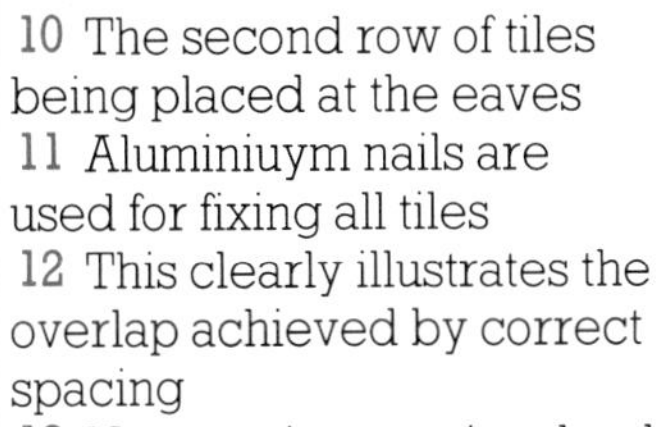

10 The second row of tiles being placed at the eaves
11 Aluminiuym nails are used for fixing all tiles
12 This clearly illustrates the overlap achieved by correct spacing
13 Neat cutting was involved to match the pitch of the roof
14 The finished job.

fixed, with 75mm overlaps horizontally, and 150mm vertically.

Batten sizes are governed by the spacing between supports, but as a guide, supports not exceeding 450mm apart need 38mm by 19mm Tanalised battens, secured with 75mm by 3.35mm galvanized nails. Spacing between battens was 115mm. You will see from the photos that a double course of tiles was laid at the eaves, formed by a course of shorter eaves tiles 190mm long with a course of full tiles laid broken bond on top. Two 38mm by 2.65mm aluminium nails were used per tile. A little cutting was needed at the gables, and to avoid the use of small triangular pieces, tiles were splay-cut in what is termed a Winchester cut. This ensures that the last tile against the main roof verge can be nailed.

Tiling chart

HEIGHT

9	3	4	4½	5½	6	6½	7½	8	9	9½	10½	11	12	12½	13	14	14½	15½	16	17	18
8½	3	3½	4½	5	5½	6½	7	7½	8½	9	10	10½	11	12	12½	13	13½	14½	15	16	16½
8	3	3½	4	4½	5½	6	6½	7½	8	8½	9	10	10½	11	12	12½	13	13½	14½	15	15½
7½	2½	3	4	4½	5	5½	6	7	7½	8	8½	9	10	10½	11	11½	12	13	13½	14	14½
7	2½	3	3½	4	5	5½	6	6½	7	7½	8	8½	9	10	10½	11	11½	12	12½	13	13½
6½	2½	3	3½	4	4½	5	5½	6	6½	7	7½	8	8½	9	9½	10	10½	11	11½	11½	12½
6	2	2½	3	3½	4	4½	5	5½	6	6½	7	7½	8	8½	9	9½	10	10½	11	11½	12
5½	2	2½	3	3½	4	4	4½	5	5½	6	6½	7	7½	7½	8	8½	9	9½	10	10½	11
5	2	2	2½	3	3½	4	4	4½	5	5½	6	6	6½	7	7½	8	8	8½	9	9½	10
4½	1½	2	2½	3	3	3½	4	4	4½	5	5½	5½	6	6½	6½	7	7½	8	8	8½	9
4	1½	2	2	2½	3	3	3½	4	4	4½	5	5	5½	5½	6	6½	6½	7	7½	7½	8
3½	1½	1½	2	2	2½	3	3	3½	3½	4	4	4½	4½	5	5½	5½	6	6	6½	6½	7
3	1	1½	1½	2	2	2½	2½	3	3	3½	3½	4	4	4½	4½	5	5	5½	5½	6	6
2½	1	1	1½	1½	2	2	2	2½	2½	3	3	3	3½	3½	4	4	4	4½	4½	5	5
2	1	1	1	1½	1½	1½	2	2	2	2½	2½	2½	3½	3	3	3½	3½	3½	4	4	4
1½	½	1	1	1	1	1½	1½	1½	1½	2	2	2	2	2½	2½	2½	2½	3	3	3	3
	2	2½	3	3½	4	4½	5	5½	6	6½	7	7½	8	8½	9	9½	10	10½	11	11½	12

LENGTH

This is the Cristal tile planning chart, and it indicates the number of tile packs required for a given wall area. The number is the same for a pack of 50 4¼in. sq. tiles as a pack of 25 6in. tiles.

The left hand vertical line shows the height of the room in feet, and the bottom horizontal line the length of the wall in feet. Treat each wall separately and read off the number of packs required. If there are any doors or windows, subtract the numberof packs relevant to the size of opening. The numbers in the squares represent the number of packs.

The table does not allow for cut tiles or additional tiles required for directional patterns

Index